Recent Topics and Innovations in Quantum Field Theory

Edited by Zbigniew Piotr Szadkowski

Published in London, United Kingdom

Recent Topics and Innovations in Quantum Field Theory
http://dx.doi.org/10.5772/intechopen.100713
Edited by Zbigniew Piotr Szadkowski

Contributors
Jishad Kumar, Vahram M. Mekhitarian, Andria Rogava, Detlef Hoyer, John W. Kenney III, Joshua Jacobsen, Amanda Renfro, Isaac Eduardo Muñoz, Ruth Christian

First published in London, United Kingdom, 2023 by IntechOpen
IntechOpen is the global imprint of INTECHOPEN LIMITED, registered in England and Wales, registration number: 11086078, 5 Princes Gate Court, London, SW7 2QJ, United Kingdom

British Library Cataloguing-in-Publication Data
A catalogue record for this book is available from the British Library

Additional hard and PDF copies can be obtained from orders@intechopen.com

Recent Topics and Innovations in Quantum Field Theory
Edited by Zbigniew Piotr Szadkowski
p. cm.
Print ISBN 978-1-80355-114-2
Online ISBN 978-1-80355-115-9
eBook (PDF) ISBN 978-1-80355-116-6

For EU product safety concerns:
IN TECH d.o.o., Prolaz Marije Krucifikse Kozulić 3, 51000 Rijeka, Croatia,
info@intechopen.com or visit our website at intechopen.com.

Meet the editor

Zbigniew Szadkowski obtained his Ph.D. in Theoretical Physics from the University of Lodz, Poland, in 1987 with his thesis, "Quark mixing in the SU(4) x SU(4) and SU(6) x SU(6) chiral symmetries." In 2007 he obtained a DSci in Experimental Physics based on the thesis, "Triggers in the Pierre Auger Observatory." Since 1998, Dr. Szadkowski has been working in the Pierre Auger Observatory designing the second-level trigger implemented in all 24 fluorescence telescopes, the first-level trigger implemented in all 1660 surface detectors, the Front End Boards with 800 ACEX and 900 Cyclone field-programmable gate arrays (FPGAs), and many trigger algorithms based on discrete cosine transform, artificial neural networks, or fuzzy logic. He worked at the Michigan Technological University, USA, the College de France, Paris, and the Bergische Universität Wuppertal, Germany.

Contents

Preface

This book discusses several topics in modern physics, including spinor fields, key outcomes of 5D relativity, dissipative quantum system and energy balance, centrifugal acceleration in relativistic astrophysics, and time-dependent photoluminescence and photoluminescence excitation in exciton systems and related phenomena.

Chapter 1, "Spinor Fields", proposes a spinor representation of the generalized energy-momentum density 4-vector. In geometry and physics, spinors are elements of a complex number-based vector space that can be associated with Euclidean space. A spinor transforms linearly when the Euclidean space is subjected to an infinitesimal rotation, but unlike geometric vectors and tensors, a spinor transforms to its negative when the space rotates through 360 degrees. It takes a rotation of 720 degrees for a spinor to go back to its original state. This property characterizes spinors. Spinors can be viewed as the "square roots" of vectors, although this is inaccurate and may be misleading. They are better viewed as "square roots" of sections of vector bundles; in the case of the exterior algebra bundle of the cotangent bundle, they thus become "square roots" of differential forms. It is also possible to associate a substantially similar notion of spinor to Minkowski space, in which case the Lorentz transformations of special relativity play the role of rotations. Spinors were introduced in geometry by Élie Cartan in 1913. In the 1920s, physicists discovered that spinors are essential to describe the intrinsic angular momentum or "spin" of the electron and other subatomic particles.

The proposed spinor representation corresponds to the classical representation of the particle's own rotation, which is described by the diagonal matrix of the moment of inertia. The concept of self-angular rotation of a particle is defined as a spatial characteristic of the field, at each point of which there is a local vortex rotation with an angular velocity Ω – a spinor field. The matrix representation of the vortex rotation Ω (spinor) and the values of the components of such a representation are derived from the matrix representation of the Lorentz transformation. The traditional concept of spin–orbit interaction, as the interaction of the magnetic moment of a particle with the magnetic field of orbital motion, is presented as the interaction of a charged particle with a spinor field. Solutions to the problems of particle motion in an external spinor field in the case of a hydrogen-like atom and planetary motion, splitting of the electron energy levels of an atom in an external magnetic field, deflection of a photon by the gravitational field, and representations in metric spaces are presented.

Chapter 2, "Key Outcomes of 5D Relativity", discusses Kaluza–Klein theory, which is general relativity extended to 5D. The gravitational constant is substituted by a scalar field, making it variable. This scalar field is predicted to change under strong dynamic electromagnetic fields. Deriving the equation of motion from this 5D metric predicts a fifth force. In physics, Kaluza–Klein theory is a classic unified field theory of gravitation and electromagnetism built around the idea of a fifth dimension beyond the common 4D of space and time and considered an important precursor to string theory. Gunnar Nordström had an earlier, similar idea. But in that case, a fifth component was added to the electromagnetic vector potential, representing the Newtonian gravitational potential and writing the Maxwell equations in five dimensions.

The 5D theory was developed in three steps. The original hypothesis came from Theodor Kaluza, who sent his results to Einstein in 1919 and published them in 1921. Kaluza presented a purely classical extension of general relativity to 5D, with a metric tensor of 15 components. Ten components are identified with the 4D space-time metric, four components with the electromagnetic vector potential, and one component with an unidentified scalar field sometimes called the "radion" or the "dilaton." Correspondingly, the 5D Einstein equations yield the 4D Einstein field equations, the Maxwell equations for the electromagnetic field, and an equation for the scalar field. Kaluza also introduced the "cylinder condition" hypothesis that no component of the 5D metric depends on the fifth dimension. Without this restriction, terms are introduced that involve derivatives of the fields with respect to the fifth coordinate, and this extra degree of freedom makes the mathematics of the fully variable 5D relativity enormously complex. Standard 4D physics seems to manifest this "cylinder condition" and, along with it, simpler mathematics.

The Pioneer 10 and 11 spacecraft were not where they were supposed to be. These missions, launched in 1972 and 1973, have covered hundreds of millions of kilometers, heading toward the edge of our solar system. But something was holding them back. Each year, they fell behind in their projected travel by about 5000 kilometers (3000 miles). In the end, after the recovery of the data and years of painstaking work, the researchers concluded that anisotropic thermal radiation (heat radiating from the spacecraft unevenly in many directions) explained the mystery of the Pioneers' deceleration. No revolution in physics was needed.

Chapter 3, "Dissipative Quantum System and Energy Balance", discusses how various parts of a quantum many-body system exchange energies at thermal equilibrium. The author assumes a quantum system is coupled to a many-body environment (at thermal equilibrium with a bigger environment) consisting of a large number of independent and non-interacting quantum harmonic oscillators above a stable ground state. Once the coupling to a large environment is switched on, the system dissipates its energy continuously to the environment until it reaches equilibrium with the latter. The author uses the Quantum Langevin equation to show such energy exchange at equilibrium and concludes that different parts of a physical system can exchange energies even at absolute zero temperature. The author computed the instantaneous power supplied by the fluctuating (random) force, which provides information about the work done by the random force on the quantum subsystem of interest. The quantum formalism is used here to verify that, at equilibrium, the work done by the fluctuating force balances the energy lost by the quantum subsystem to the heat bath. The quantum subsystem chosen to couple to the heat bath is the charged oscillator in a magnetic field. The calculations were done using the Drude regularized spectral density of bath oscillators instead of using a strict ohmic spectral density that gives memoryless damping.

Chapter 4, "Centrifugal Acceleration in Relativistic Astrophysics", is devoted to the centrifugal acceleration of astro-particles to relativistic energies that might take place in rotating astrophysical objects. It is strongly believed that active galactic nuclei and pulsars have rotating magnetospheres; therefore, they potentially can drive charged particles to high and ultra-high energies. It is a proposed explanation for ultra-high-energy cosmic rays (UHECRs) and extreme-energy cosmic rays (EECRs) exceeding the Greisen–Zatsepin–Kuzmin limit. The author gives a brief review of the theoretical ideas and results related to Machabeli and Rogava gedankenexperiment (1994) and their astrophysical implications. In particular, three astrophysical cases, including acceleration of particles by rotating magnetospheres in AGNs, centrifugal acceleration and gamma flares in Crab nebula, and self-trapping as a beaming mechanism for Fast Radio Bursts, are discussed.

Although direct centrifugal acceleration has limitations, analysis shows the effects of rotation still might play an important role in the processes of acceleration of charged particles. Generally speaking, it is believed that the centrifugal relativistic effects may induce plasma waves, which under certain conditions might be unstable efficiently pumping energy from the background flow. In the second stage, the energy of wave modes can be transformed into the energy of plasma particles, leading to consequent acceleration. In rotating magnetospheres, the centrifugal force acts differently in different locations, leading to the generation of Langmuir waves or plasma oscillations via the parametric instability. One can show that this mechanism efficiently works in the magnetospheres of AGN and pulsars.

Chapter 5 is titled "Time-Dependent Photoluminescence and Photoluminescence Excitation in Exciton Systems and Related Phenomena". When an electron is excited into a higher energy state, either through absorption of a photon or another excitation method (such as in electroluminescence), this creates a positively charged space in the lower energy level known as a "hole." This results in the formation of an electron–hole pair. In some cases, these two particles exist in a bound state, forming a single quasi-particle known as an exciton. Within an exciton, the electron and hole are bound together by Coulombic interactions, and the strength of this bond is quantified by its exciton binding energy. Excitons can be grouped into two categories: Frenkel excitons and Wannier–Mott excitons. Frenkel excitons are tightly bound excitons that have a radius similar in magnitude to the crystal unit cell or atomic radii. These are often found in insulators or organic semiconductors, as they are often bound to specific atoms or molecules. Alternatively, Wannier–Mott excitons have a larger excitonic radius. The exciton therefore encompasses many unit cells and thus can move more freely through the crystal structure. These are commonly found in inorganic semiconductors, and their excitonic bonds can be overcome at room temperature. Generally, organic matter such as fullerenes and insulators tend to exhibit Frenkel excitons with a larger binding energy. These are sometimes known as bound electrons. Most inorganic or hybrid semiconductors (such as perovskites) have Wannier–Mott excitons and free excitons that can travel through the crystal structure. Investigations of exciton systems represent one of the fastest-growing, multifaceted, and productive areas of modern solid-state physics and material science. Particularly noteworthy in this regard is the deployment of state-of-the-art spectroscopic techniques (e.g., tr-ARPES and femtosecond optical spectroscopy) to better elucidate the structural, dynamical, and quantum mechanical properties of exciton systems.

Selected topics presented in the chapters come from very different fields, including theoretical physics, astrophysics, and solid-state physics. The analysis of the chapters requires the reader to have very advanced knowledge of mathematics, theoretical physics, and so on. However, even a reader unfamiliar with the topic can learn about new directions or tendencies in contemporary physics.

Zbigniew Piotr Szadkowski
Faculty of Physics and Applied Informatics,
Department of Intelligent Systems,
University of Lodz,
Lodz, Poland

Chapter 1

Spinor Fields

Vahram Mekhitarian

Abstract

A spinor representation of the generalized energy-momentum density 4-vector is proposed, and examples of such representations for various particles and fields are given. This representation corresponds to the classical representation of the particle's own rotation, which is described by the diagonal matrix of the moment of inertia. The concept of self-angular rotation of a particle is defined as a spatial characteristic of the field, at each point of which there is a local vortex rotation with an angular velocity $\mathbf{\Omega}$ – a spinor field. The matrix representation of the vortex rotation $\mathbf{\Omega}$ (spinor) and the values of the components of such a representation are derived from the matrix representation of the Lorentz transformation. The traditional concept of spin-orbit interaction, as the interaction of the magnetic moment of a particle with the magnetic field of orbital motion, is presented as the interaction of a charged particle with a spinor field. Solutions to the problems of particle motion in an external spinor field in the case of a hydrogen-like atom and planetary motion, splitting of the electron energy levels of an atom in an external magnetic field, deflection of a photon by the gravitational field, and representations in metric spaces are presented.

Keywords: spinor fields, relativistic and quantum mechanics, spin concept, Lorentz transformation, spinor fields representation

1. Introduction

When I meet God, I am going to ask Him two
questions: Why relativity? And why turbulence?
I really believe He will have an answer for the first.

Attributed to Werner Heisenberg (1901–1976).

The material presented in this chapter is based on the new approaches of relativistic and quantum mechanics developed in the works [1–4]. Equations, which are obtained by applying the invariance principle for the total four-dimensional momentum of the system "field + particle," have some significant advantages as compared with its analog equations such as the Klein-Fock-Gordon and Dirac equations. For instance, the problem of a hydrogen-like atom has solutions for an arbitrary value of the interaction constant not restricted to whatever the atomic number of the nucleus (we recall that for the Dirac equation the atomic number is restricted to $Z < 137$).

In contrast to the well-known equations of relativistic and quantum mechanics, the energy levels of the ground state of the particle for the considered equation prove to be limited by the size of the spatial characteristic. This property directly reflects the uncertainty principle in that, irrespective of the well depth value, the particle can be localized in a bound state only if the well width is larger than the half-wavelength of the particle.

For the problem of the passage of a particle through a potential barrier, if the energy of the particle does not exceed the height of the potential barrier, then the transmission coefficient is equal to zero regardless of the height of the barrier. In this case, there is no contradiction like Klein's paradox.

The equations are applicable for different types of particles and interactions. The analysis of the solutions shows full compliance with the principles of relativistic and quantum mechanics, and the solutions are devoid of any restrictions on the nature and magnitude of the interactions.

Representation of the generalized momentum	
$\mathrm{P} = \begin{pmatrix} \frac{\mathrm{mc}}{\sqrt{1-\beta^2}} + \frac{q}{c}\varphi, \\ \frac{\mathrm{mc}}{\sqrt{1-\beta^2}}\boldsymbol{\beta} + \frac{q}{c}\mathrm{A} \end{pmatrix}$ $\mathrm{P}^2 = \boldsymbol{\varepsilon}^2 - \mathrm{p}^2 = \boldsymbol{I}^2 \neq \text{invariant}$	$\mathrm{P} = \begin{pmatrix} \frac{\mathrm{mc}}{\sqrt{1-\beta^2}} + \frac{q}{c}\frac{\varphi + \boldsymbol{\beta}\cdot\mathrm{A}}{\sqrt{1-\beta^2}}, \\ \frac{\mathrm{mc}}{\sqrt{1-\beta^2}}\boldsymbol{\beta} + \frac{q}{c}\frac{\varphi\boldsymbol{\beta} + \mathrm{A}_{\parallel}}{\sqrt{1-\beta^2}} + \frac{q}{c}\mathrm{A}_{\perp} \end{pmatrix}$ $\mathrm{P}^2 = \boldsymbol{\varepsilon}^2 - \mathrm{p}^2 = \boldsymbol{I}^2 = \text{invariant}$
Invariant of the generalized momentum	
$\mathbf{P}^2 = \left(\varepsilon - \frac{q}{c}\varphi\right)^2 - \left(\mathbf{p} - \frac{q}{c}\mathbf{A}\right)^2 = (mc)^2$	$\mathbf{P}^2 = \varepsilon^2 - \mathbf{p}^2 = \left(mc + \frac{q}{c}\varphi\right)^2 - \left(\frac{q}{c}\mathbf{A}\right)^2$
Space-time interval	
$ds^2 = I^2\left(d\tau^2 - dx^2 - dy^2 - dz^2\right) = g_{ik}dx^i dx^k$ $I^2 = (mc)^2$	$ds^2 = I^2\left(d\tau^2 - dx^2 - dy^2 - dz^2\right) = g_{ik}dx^i dx^k$ $I^2 = \left(mc + \frac{q}{c}\varphi\right)^2 - \left(\frac{q}{c}\mathbf{A}\right)^2$
Hamilton-Jacobi relativistic equation	
$\left(\frac{\partial S}{\partial \tau} - \frac{q}{c}\varphi\right)^2 - \left(-\frac{\partial S}{\partial \mathbf{r}} - \frac{q}{c}\mathbf{A}\right)^2 = (mc)^2$	$\left(\frac{\partial S}{\partial \tau}\right)^2 - \left(\frac{\partial S}{\partial \mathbf{r}}\right)^2 = \left(mc + \frac{q}{c}\varphi\right)^2 - \left(\frac{q}{c}\mathbf{A}\right)^2$
Hamilton-Jacobi-Einstein equation	
$g_{ik}\sqrt{-g}\left(\frac{\partial S}{\partial x^i} - A_i\right)\left(\frac{\partial S}{\partial x^k} - A_k\right) = (mc)^2$	$g_{ik}\sqrt{-g}\,\frac{\partial S}{\partial x^i}\,\frac{\partial S}{\partial x^k} = \left(mc + \frac{q}{c}\varphi\right)^2 - \left(\frac{q}{c}\mathbf{A}\right)^2$
Klein-Fock-Gordon equation	
$\left(i\hbar\frac{\partial}{\partial \tau} - \frac{q}{c}\varphi\right)^2\Psi - \left(-i\hbar\frac{\partial}{\partial \mathbf{r}} - \frac{q}{c}\mathbf{A}\right)^2\psi =$ $-(mc)^2\psi$	$\frac{\partial^2\psi}{\partial\tau^2} - \frac{\partial^2\psi}{\partial\mathbf{r}^2} =$ $-\frac{\left(mc^2 + q\varphi\right)^2 - (q\mathbf{A})^2}{\hbar^2c^2}\psi$
Klein-Fock-Gordon equation in metric spaces	
$\frac{1}{\sqrt{-g}}\left(i\hbar\frac{\partial}{\partial x^i} - \frac{q}{c}A_i\right)\times$ $\left(g_{ik}\sqrt{-g}\left(-i\hbar\frac{\partial}{\partial x^k} - \frac{q}{c}A_k\right)\psi\right) = -(mc)^2\psi$	$\frac{1}{\sqrt{-g}}\frac{\partial}{\partial x^i}\left(g_{ik}\sqrt{-g}\frac{\partial\psi}{\partial x^k}\right) =$ $-\frac{\left(mc^2 + q\varphi\right)^2 - (q\mathbf{A})^2}{\hbar^2c^2}\psi$
Dirac equation in 1/2 spin spaces	
$\left(\hat{\varepsilon} - \frac{q}{c}\varphi\right)\phi - \boldsymbol{\sigma}\cdot\left(\hat{\mathbf{p}} - \frac{q}{c}\mathbf{A}\right)\chi = mc\phi,$ $\left(\hat{\varepsilon} - \frac{q}{c}\varphi\right)\chi + \boldsymbol{\sigma}\cdot\left(\hat{\mathbf{p}} - \frac{q}{c}\mathbf{A}\right)\phi = -mc\chi.$	$\hat{\varepsilon}\phi - \boldsymbol{\sigma}\cdot\hat{\mathbf{p}}\,\chi = \left(mc + \frac{q}{c}\varphi\right)\phi - \frac{q}{c}\boldsymbol{\sigma}\cdot\mathbf{A}\chi,$ $\hat{\varepsilon}\chi + \boldsymbol{\sigma}\cdot\hat{\mathbf{p}}\,\phi = -\left(mc + \frac{q}{c}\varphi\right)\chi - \frac{q}{c}\boldsymbol{\sigma}\cdot\mathbf{A}\phi.$

Table 1.
Old and new [1–4] concepts and equations of relativistic and quantum mechanics.

However, this new theory was presented without an explicit representation of the spinor properties of fields and systems (**Table 1**). In this chapter, based on the representation of local relativistic rotation by the Lorentz transformation matrix, the representations of the four-dimensional energy-momentum vector for various spinor fields and systems, the results of corresponding solutions of the new equations of relativistic and quantum mechanics are given.

Since within the framework of the new, generalized relativistic theory, there is an exact correspondence between the representations of relativistic, quantum mechanics, and general relativity, in the chapter spinor properties and equations are also presented in metric spaces.

2. The spin concept

Although in physics the concept of spin arose as a property of the proper rotation of a particle (electron, G. E. Uhlenbeck, S. Goudsmit), as a result of quantization of the self-angular momentum, further development by W. Pauli and P. Dirac led to the description of spin as a property of space itself, in which we describe particles. Their interactions corresponded to the data of the physical experiment, when they were presented in spinor spaces.

One of the first, the Dirac equation, which describes systems in spaces with spin 1/2, the solution in the case of the hydrogen atom gives a very good match with the real spectrum of the hydrogen atom. But the hydrogen atom problem (**Figure 1**),

which consists of a proton and an electron, is solved for a generalized particle with the reduced mass m of an electron m_1 and a proton m_2 (the two-body problem), whose coordinates $\mathbf{r}$ do not coincide with either the coordinates of the electron $\mathbf{r}_1$ or the coordinates of the proton $\mathbf{r}_2$, and the spin of this generalized particle (the sum of the spins of the proton and electron) can only have values of 0 or 1.

Obviously, Dirac's spin 1/2 refers to the properties of the space in which the hydrogen atom is described, not to an electron, a proton, or a hydrogen atom. Similarly, regardless of the spin of the nucleus and the spin of the electron shell of hydrogen-like atoms, the spectrum, and fine splitting is described by solving the Dirac equation with spin 1/2. The same is true for other problems, regardless of the properties of the components of the physical system themselves—solutions to the Dirac equations describe systems with only a spin of 1/2.

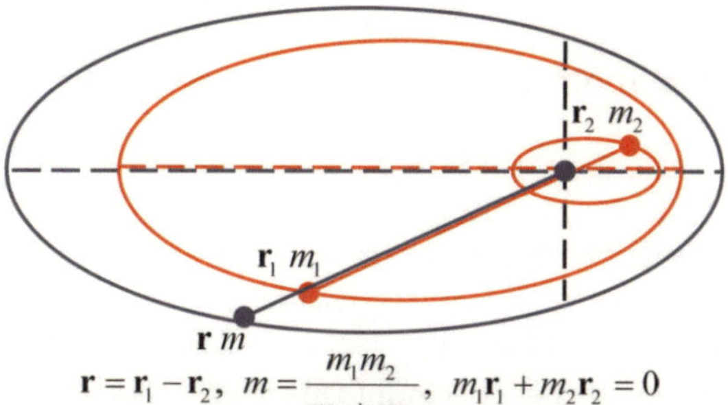

$$\mathbf{r} = \mathbf{r}_1 - \mathbf{r}_2, \quad m = \frac{m_1 m_2}{m_1 + m_2}, \quad m_1\mathbf{r}_1 + m_2\mathbf{r}_2 = 0$$

Figure 1.
The problem of the motion of two bodies can be represented as a problem of the motion of one body.

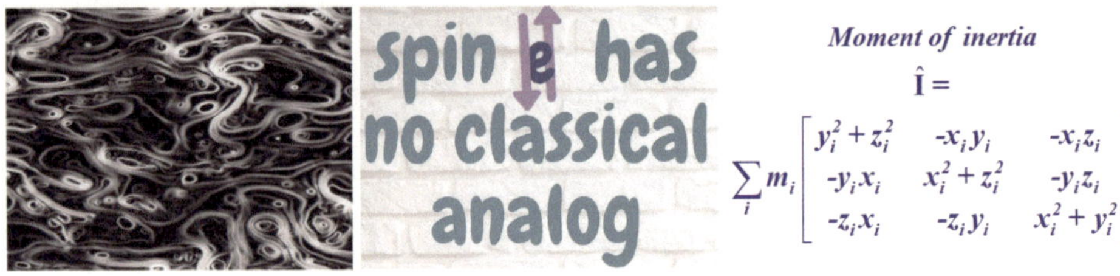

Figure 2.
The concept of spin in quantum mechanics does not describe the physical properties of the vortex motion or proper rotation of bodies in the classical sense.

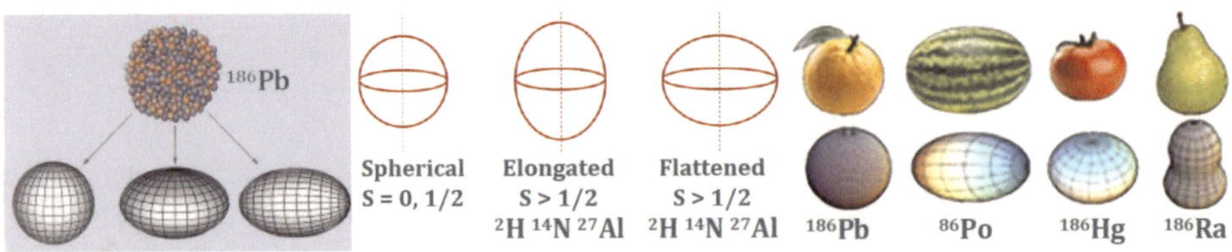

Figure 3.
The projections of the self-angular momentum (spin) of nonspherical nuclei have different values in the directions of the main axes of inertia.

Naturally, the spinor properties of space do not in any way describe the physical properties of the self-angular momentum of a particle or a vortex (turbulent) field (the flow of a liquid or gas) in the classical sense, and therefore, the statement arose that spin has no classical analog (**Figure 2**). And the problem of describing the self-angular momentum of the particles and the vortex field remained unresolved.

Note that the description of the motion of an asymmetric spinning top, represented by the tensor of the moment of inertia, has no analog in quantum mechanics since the modern concept of spin does not imply any representations of the concept of the moment of inertia (**Figure 2**). But in the case of nonspherical nuclei (**Figure 3**), the projection of the self-angular momentum must have different spin values relative to the main axes of inertia (rotation) of the nucleus.

The main, fundamental physical variable, for which the variational principles and equations of relativistic and quantum mechanics are formulated, is the energy-momentum density 4-vector, so all the properties of the system, also spinor, must be initially reflected in the representation of the energy-momentum 4-vector $\mathbf{P} = (\varepsilon, \mathbf{p})$.

3. Orbital and vortex motions of the continuous media

Imagine holding a bicycle wheel by the axle at a distance l and rotating around its axis with an angular velocity of $\boldsymbol{\Omega}$ (**Figure 4**).

Since the wheel rotates freely around its axis, there is no moment of own rotation of the wheel itself, and it makes an only translational motion (movement without its own rotation). Energy W is expressed by the usual formula of kinetic energy through the mass m and velocity $\mathbf{v}$ of the wheel movement $W = m\mathbf{v}^2/2 = ml^2\boldsymbol{\Omega}^2/2$.

If there is any friction of the axle, then the rotational motion of the axle will gradually be transmitted to the wheel and eventually, the wheel will rotate with the same angular speed. Note that in this case, we considered the option of a rigidly fixed

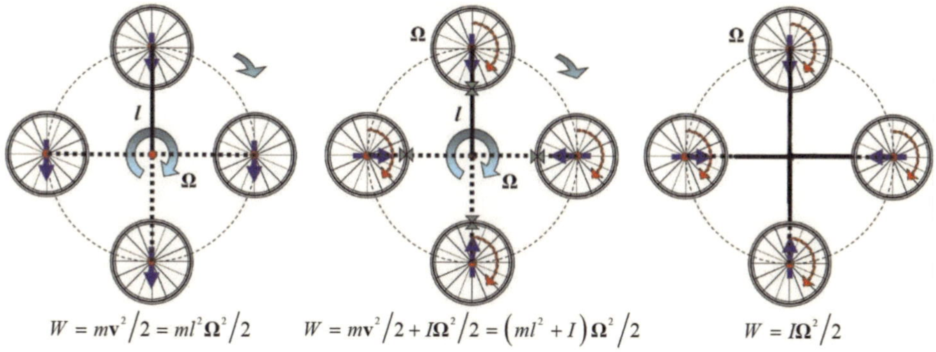

Figure 4.
Bicycle wheel at a distance l rotating around axis with an angular velocity of $\mathbf{\Omega}$*.*

axis (rigidly fixed center of mass) wheel. This corresponds to a rigid spin-orbit interaction when the angular velocities of rotation are the same.

In this case, the energy is equal to the sum of the kinetic energy of the wheel and the energy of the wheel's own rotation with the moment of inertia I (a sample conclusion of the Huygens-Stern theorem)

$$W = m\mathbf{v}^2/2 + I\mathbf{\Omega}^2/2 = \left(ml^2 + I\right)\mathbf{\Omega}^2/2. \tag{1}$$

In other cases, in terms of energy, there will be other ratios of translational energy and energy of its own rotation (spin). In particular, the motion of the Moon corresponds to the case of a rigidly fixed center of mass (the Moon is constantly facing the Earth on the same side), and the movement of the Earth around the Sun corresponds to the case of a freely fixed center of mass (the tilt of the axis, the period of its own rotation is in no way related to the movement around the Sun).

Note that regardless of the distance from the center of rotation, the angular velocity of its own rotation and energy is constant. If the system is represented as a medium with distributed local rotation, then it can be described by the energy density of vortex rotation with an angular velocity $\mathbf{\Omega}$. A good example of such a system is a permanent magnet, where at each point of the medium there are eddy currents (rotating electric field) and, accordingly, the magnetic field of these currents (rotating electric field).

To illustrate the vortex motion of the distributed systems, let us consider an example of a large, thin hard disk on which much identical small metal (heavy) spinners of mass m and moment of inertia I are attached perpendicularly and densely (**Figure 5**). The mass of the disk relative to the total mass of spinners Σm can be neglected. If the spinners do not have axial friction, then when the disk rotates, the spinners will rotate around the axis of the disk, but will not rotate around their own axis—the arrows will always be turned toward the original direction (**Figure 5**). In this case, the rotation energy of the system W is determined only by the sum of the kinetic energies of the orbital rotation of the spinners $W = \Sigma m\mathbf{v}_i{}^2/2 = \Sigma m r_i{}^2\mathbf{\Omega}^2/2$ relative to the axis of the disk, and when the disk stops, the energy of the system becomes zero.

In the case when the spinners have some small axial friction (like the spin-orbit interaction of an electron in an atom or the connection of the moon's rotation with the Earth's rotation due to tides), then when the disk rotates, due to friction, after some

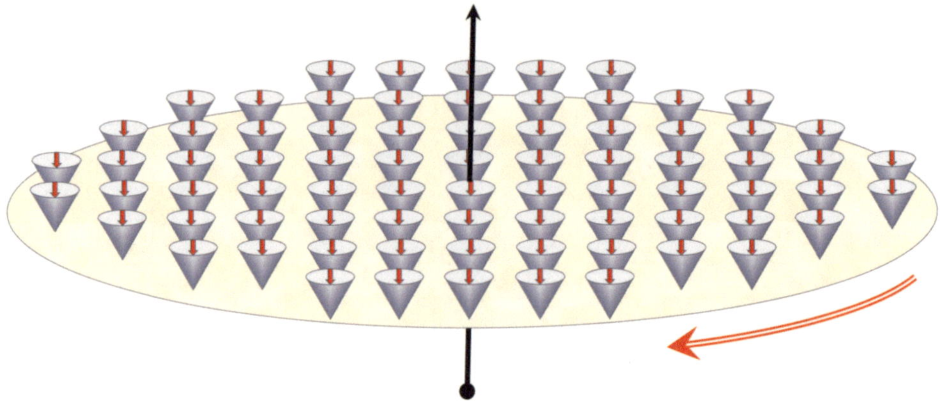

Figure 5.
Spinners of mass m *and moment of inertia* I *are perpendicularly and densely fixed on the thin hard disk.*

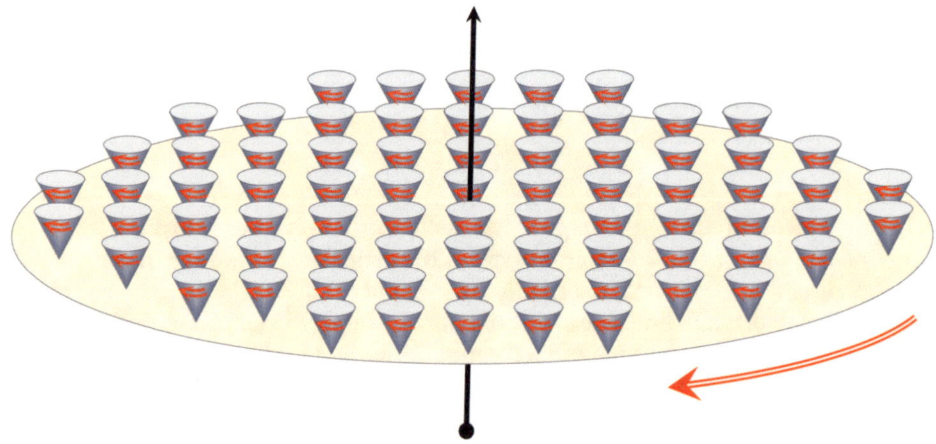

Figure 6.
Spinners rotate around their axis with the angular velocity of the disk **Ω**.

time the spinners will begin to rotate around its own axis and with the angular speed of rotation of the disk **Ω** (**Figure 6**). In this case, during rotation, the spinners are at any moment oriented toward the center of the disk axis (like a Moon), which corresponds to the rotation of the disk with the axes of the spinners fixed rigidly—at each point, the spinners rotate around their own axis with the angular speed of rotation of the disk. The energy of the system will be represented as the sum of the kinetic energy of the orbital rotation and the energy of the own rotation of the spinners in the form of $W = \Sigma m{r_i}^2\mathbf{\Omega}^2/2 + \Sigma I\mathbf{\Omega}^2/2$.

If we stop the rotation of the disk in such an established stationary state, then the energy of the orbital motion of the spinners will be reset, but the energy of its own rotation $\Sigma I\mathbf{\Omega}^2/2$ will be preserved (**Figure 7**).

The picture, obtained after stopping the disk, represents a distributed system with local vortex motion. Regardless of the distance from the center of the disk, the angular velocity of rotation and energy of the spinners are constant. If the system is represented as a medium with distributed local rotation, then it can be described by the energy density of vortex rotation with an angular velocity **Ω**.

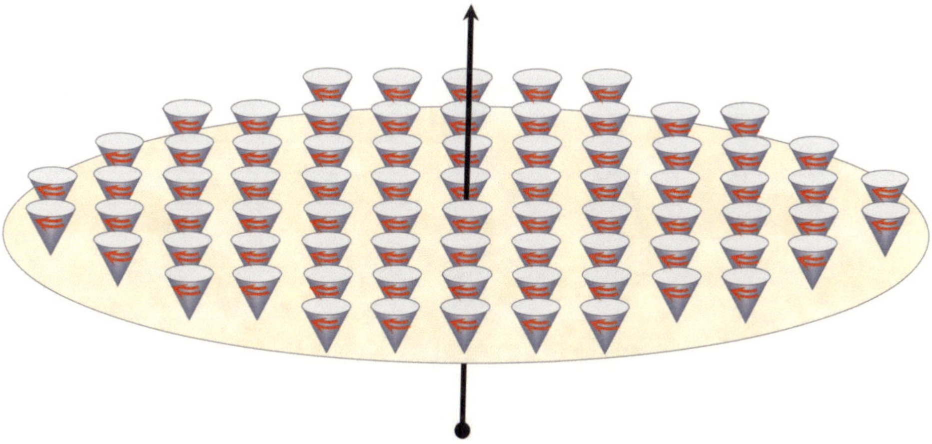

Figure 7.
The spinners keep spinning after the disk has stopped.

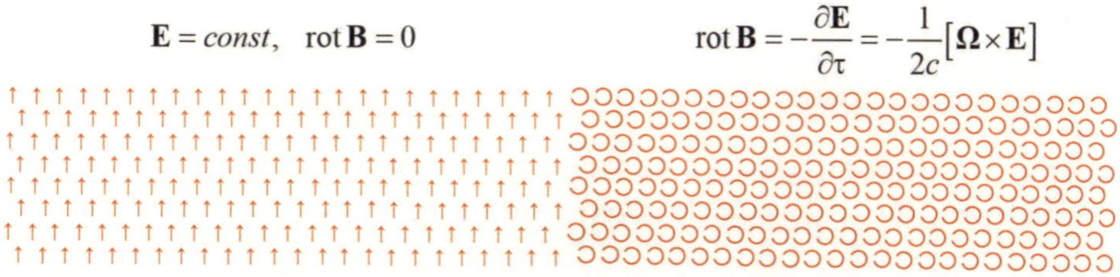

Figure 8.
Eddy currents and magnetic field exist at each point of the medium.

Let us emphasize that although the described system does not have orbital momentum, the generalized angular momentum and the total energy of the system are not zero and have minimal internal angular momentum and energy. It should be noted that in the ground state, both the angular momentum and the energy of the system have corresponding minimum values due to the internal rotation—spin.

A good example of such a system is a permanent magnet, where eddy currents (rotating electric field) and, accordingly, the magnetic field of these currents (rotating electric field) exist at each point of the medium (**Figure 8**).

Vortex and circular fields should be distinguished: in vortex fields, the rotor is nonzero at any point in the field, and in circular fields, it is zero (**Figure 9**). Such is the electric field outside the alternating current solenoid, where the magnetic field is zero.

A time-varying magnetic field in the solenoid generates, induces an electric field **E**, which is described by Maxwell's equations. Such a field is described in cylindrical coordinates and is represented by the vector potential $\mathbf{A} = \mathbf{A}(\mathbf{r}, \ t)$ in the form [5].

$$\mathbf{A}(\mathbf{r}, t) = \begin{cases} \frac{1}{2}[\mathbf{B}(t) \times \mathbf{r}], & r \le R_S; \\ \frac{{R_S}^2}{2(x^2 + y^2)}[\mathbf{B}(t) \times \mathbf{r}], & r > R_S; \end{cases} \tag{2}$$

where R_S is a radius of the solenoid.

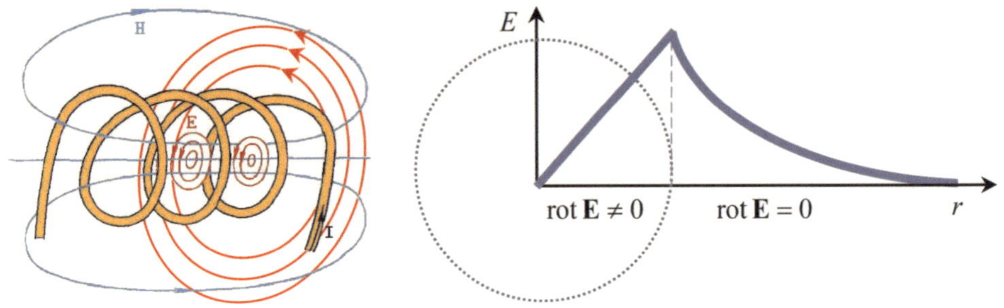

Figure 9.
Vortex and circular fields of the solenoid.

For the fields of the solenoid, we have

$$\mathbf{B} = \text{rot}\mathbf{A} = \begin{cases} \mathbf{B}, & r \le R_S, \\ 0, & r > R_S. \end{cases} \qquad \mathbf{E} = -\frac{1}{c}\frac{\partial \mathbf{A}}{\partial t} = \begin{cases} \dfrac{1}{2c}\left[\mathbf{r} \times \dfrac{\partial \mathbf{B}}{\partial t}\right], & r \le R_S, \\ \dfrac{{R_S}^2}{2c(x^2+y^2)}\left[\mathbf{r} \times \dfrac{\partial \mathbf{B}}{\partial t}\right], & r > R_S. \end{cases} \tag{3}$$

Inside the solenoid, the electric field is vortex – rot$\mathbf{E} \neq 0$, and outside the solenoid it is circular and the magnetic field is zero – rot$\mathbf{E} = 0$.

4. Representation of the spinor fields in Minkowski spaces

When representing continuously distributed systems, one should average the energy of the spinners over the occupied volume and describe the continuous medium with the energy-momentum distribution density.

If we want to describe the spatial properties of the vortex fields, then rotation angular velocity $\boldsymbol{\Omega}$ at a given point in space can serve as such a kinematic variable. The occurrence of vortex rotation distributed in space in the framework of the relativistic theory is described by the Lorentz transformation, calculating the energy-momentum distribution density of local rotations.

For fields, we have [6]

$$\begin{aligned}(\varphi', \ \mathbf{A}') &= \left(\frac{\varphi + \boldsymbol{\beta}\cdot\mathbf{A}}{\sqrt{1-\beta^2}}, \ \frac{\varphi\boldsymbol{\beta} + q\mathbf{A}_\parallel}{\sqrt{1-\beta^2}} + q\mathbf{A}_\perp\right) = \\ &(\varphi, \ \mathbf{A}) + \left((\gamma-1)\varphi + \gamma\boldsymbol{\beta}\cdot\mathbf{A}, \quad \gamma\varphi\boldsymbol{\beta} + (\gamma-1)(\mathbf{A}\cdot\boldsymbol{\beta})\boldsymbol{\beta}/\beta^2\right).\end{aligned} \tag{4}$$

where $\gamma = 1/\sqrt{1-\beta^2}$ is a Lorentz factor. This transformation can be presented in matrices form

$$(\varphi', \quad \mathbf{A}') = (\varphi, \quad \mathbf{A}) + \hat{\mathrm{T}} \ (\varphi, \quad \mathbf{A}). \tag{5}$$

where a Lorentz transformation has a form [1].

$$\hat{\mathrm{T}} = \gamma \begin{Vmatrix} 0 & \beta_1 & \beta_2 & \beta_3 \\ \beta_1 & 0 & 0 & 0 \\ \beta_2 & 0 & 0 & 0 \\ \beta_3 & 0 & 0 & 0 \end{Vmatrix} + (\gamma - 1) \begin{Vmatrix} 1 & 0 & 0 & 0 \\ 0 & \frac{\beta_1\beta_1}{\beta^2} & \frac{\beta_1\beta_2}{\beta^2} & \frac{\beta_1\beta_3}{\beta^2} \\ 0 & \frac{\beta_2\beta_1}{\beta^2} & \frac{\beta_2\beta_2}{\beta^2} & \frac{\beta_2\beta_3}{\beta^2} \\ 0 & \frac{\beta_3\beta_1}{\beta^2} & \frac{\beta_3\beta_2}{\beta^2} & \frac{\beta_3\beta_3}{\beta^2} \end{Vmatrix}. \tag{6}$$

The matrices of the invariant representation of a four-dimensional vector, which preserves the vector module in four-dimensional space, form the Poincare group (*inhomogeneous Lorentz group*). In addition to displacements and rotations, the group contains space-time reflection representations $\hat{\mathrm{P}}$, $\hat{\mathrm{T}}$ and inversion$\hat{\mathrm{P}}\hat{\mathrm{T}} = \hat{\mathrm{I}}$.

In the case of a vortex field, each point can be attributed to a local rotation with an angular velocity $\mathbf{\Omega}$ and velocity $\mathbf{v} = [\mathbf{\Omega} \times \mathbf{r}]$ at a distance r from the chosen point of rotation. First, we consider rotation only around one axis of the coordinate system $\hat{\mathbf{z}}$. Then, when choosing a cylindrical coordinate system, we have

$$\gamma = 1/\sqrt{1-(v/c)^2} = 1/\sqrt{1-[\mathbf{\Omega}\times\mathbf{r}]^2/c^2} = 1/\sqrt{1-(r/r_h)^2}, \tag{7}$$

$$\hat{\mathrm{T}} = \gamma\beta \begin{Vmatrix} 0 & \cos(\phi) & \sin(\phi) & 0 \\ \cos(\phi) & 0 & 0 & 0 \\ \sin(\phi) & 0 & 0 & 0 \\ 0 & 0 & 0 & 0 \end{Vmatrix} + (\gamma - 1) \begin{Vmatrix} 1 & 0 & 0 & 0 \\ 0 & \cos^2(\phi) & \sin(\phi)\cos(\phi) & 0 \\ 0 & \sin(\phi)\cos(\phi) & \sin^2(\phi) & 0 \\ 0 & 0 & 0 & 0 \end{Vmatrix}, \tag{8}$$

were $r_h = c/\Omega$ the event horizon radius. At each point in space, we must average the value of the energy-momentum and assign it to the selected point. From (7), (8) it follows that when averaging in volume $\pi r_h{}^2 d$ ($d$is the thickness of the disk), only the average values of the diagonal elements of the matrix are nonzero:

$$\hat{\bar{\mathrm{T}}} = \frac{1}{V}\int_V \hat{\mathrm{T}} dv =$$
$$\frac{1}{\pi r_h{}^2 d}\int_0^d dz \cdot \int_0^{r_h} r\left(\frac{1}{\sqrt{1-r^2/r_h{}^2}} - 1\right) dr \cdot \int_0^{2\pi} \begin{Vmatrix} 1 & 0 & 0 \\ 0 & \cos^2(\phi) & \sin(\phi)\cos(\phi) \\ \sin(\phi)\cos(\phi) & 0 & \sin^2(\phi) \end{Vmatrix} d\phi =$$
$$\begin{Vmatrix} 1 & 0 & 0 \\ 0 & 1/2 & 0 \\ 0 & 0 & 1/2 \end{Vmatrix}, \tag{9}$$

where

$$\frac{2\pi}{\pi r_h^2}\int_0^{r_h} r\left(\frac{1}{\sqrt{1-r^2/r_h^2}}-1\right)dr=\int_0^1\left(\frac{1}{\sqrt{1-x^2}}-1\right)2xdx=-2\sqrt{1-x^2}-x^2\Big|_0^1=1. \tag{10}$$

Averaging over the entire volume to the event horizon r_h, we obtain a result independent of the speed of angular rotation **Ω** This seemingly unexpected result is due to the inverse dependence of the event horizon $r_h = c/\Omega$ on the angular velocity of rotation. The greater the speed of angular rotation, the greater the relativistic compression and, accordingly, the increase in energy density due to a decrease in the volume of integration, so the integral does not depend on the angular velocity **Ω** This means that own rotation, spin is an invariant for any reference frames, in the approaches of general relativity.

Based on the principle of superposition and additivity, such elementary excitations of the generating field in a unit of volume can be any integer, so that in the general case n

$$\hat{\boldsymbol{\Omega}}=\left\|\begin{matrix}1/2 & 0 & 0\\ 0 & 1/2 & 0\\ 0 & 0 & 0\end{matrix}\right\|_{\hat{\mathbf{z}}}\quad\rightarrow\quad\pm\frac{1}{2}\times n=s_z\quad\rightarrow\quad\left\|\begin{matrix}s_z & 0 & 0\\ 0 & s_z & 0\\ 0 & 0 & 0\end{matrix}\right\|_{\hat{\mathbf{z}}}, \tag{11}$$

where is a $\hat{\boldsymbol{\Omega}}$ representation matrix of the rotation (spinor), and $\Omega_x = \Omega_y = s_z$ – the value of components for the $\hat{\mathbf{x}}$, $\hat{\mathbf{y}}$ axis's when rotating around an axis $\hat{\mathbf{z}}$. As we see, the rotation along the selected axis generates two equal perpendicular spatial components of the generalized momentum with half-integer coefficients (**Figure 10**).

Accordingly, taking into account the independent rotations in all axes, for the matrix of the spinor representation we have

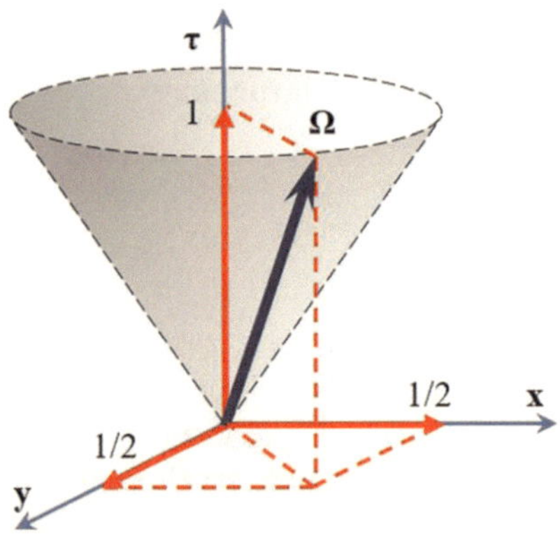

Figure 10.
Rotation along the selected axis generates two equal perpendicular spatial components with half-integer coefficients.

$$\hat{\mathbf{\Omega}} = \left\| \begin{matrix} 0 & 0 & 0 \\ 0 & s_x & 0 \\ 0 & 0 & s_x \end{matrix} \right\|_{\hat{\mathbf{x}}} + \left\| \begin{matrix} s_y & 0 & 0 \\ 0 & 0 & 0 \\ 0 & 0 & s_y \end{matrix} \right\|_{\hat{\mathbf{y}}} + \left\| \begin{matrix} s_z & 0 & 0 \\ 0 & s_z & 0 \\ 0 & 0 & 0 \end{matrix} \right\|_{\hat{\mathbf{z}}} = \left\| \begin{matrix} s_y + s_z & 0 & 0 \\ 0 & s_x + s_z & 0 \\ 0 & 0 & s_x + s_y \end{matrix} \right\|. \tag{12}$$

Thus, the spinor, the averaged Lorentz transformation for local rotations, is represented by a diagonal matrix or column as

$$\begin{gathered} \hat{\mathbf{\Omega}} = \{s_y + s_z,\ s_x + s_z,\ s_x + s_y\},\quad s_x,\ s_y,\ s_z = 0,\ \pm\frac{1}{2},\ \pm 1,\ \pm\frac{3}{2},\ \dots,\ \pm\frac{n}{2},\ \dots \\ \hat{\mathbf{\Omega}}^2 = 2\left(s_x{}^2 + s_y{}^2 + s_z{}^2 + s_x s_y + s_x s_z + s_y s_z\right). \end{gathered} \tag{13}$$

The index symbols of the components of the matrix are specifically selected to indicate from rotation around which axes these components originated:

$$\Omega_x = s_y + s_z,\quad \Omega_y = s_x + s_z,\quad \Omega_z = s_x + s_y. \tag{14}$$

Negative signs of the projections of the spinor matrix $\hat{\mathbf{\Omega}}$ appeared during the transformation of reflections and inversions (antiparticles).

$$\begin{gathered} \hat{\mathbf{T}} = \left\| \begin{matrix} \pm m & 0 & 0 & 0 \\ 0 & s_y + s_z & 0 & 0 \\ 0 & 0 & s_x + s_z & 0 \\ 0 & 0 & 0 & s_x + s_y \end{matrix} \right\|, \\ \mathbf{T} = \left(\pm m,\quad s_y + s_z,\quad s_x + s_z,\quad s_x + s_y\right). \end{gathered} \tag{15}$$

where m is integer.

If a scalar field φ is given in space, at each point of which the vortices are excited, then the vector potential of such a field can be represented as $\mathbf{A} = \varphi\mathbf{\Omega}$, and the generalized energy-momentum $\mathbf{P}$ as (**Table 2**)

$$\begin{gathered} \mathbf{P} = (\varphi,\ \mathbf{A}) = (\varphi,\ \varphi\mathbf{\Omega}),\quad \mathbf{P}^2 = \varphi^2\left(1 - \mathbf{\Omega}^2\right), \\ 1 - \mathbf{\Omega}^2 = 1 - 2\left(s_x{}^2 + s_y{}^2 + s_z{}^2 + s_x s_y + s_x s_z + s_y s_z\right). \end{gathered} \tag{16}$$

If for qualitative evaluation, we assume that the energy of such elementary excitation corresponds to the quantum of the rotator energy $E = mc^2 = \hbar\Omega$, then we get

$$\lambda\!\!\!^{-} = \hbar/mc = c/\Omega = r_h, \tag{17}$$

that is, the event horizon radius ("particle size") is the wavelength of the particle $r_h = \lambda\!\!\!^{-}$.

Since the spin of the field has been determined locally, its direction from point to point can change, provided that the internal structure is preserved. The spinor field, as in the general case of any vector potential $\mathbf{A} = \varphi\mathbf{\Omega}$, must satisfy the basic equations for the fields [2].

$(s_y + s_z,\ s_x + s_z,\ s_x + s_y)$	Ω^2	$1 - \Omega^2$	$(s_y + s_z,\ s_x + s_z,\ s_x + s_y)$	Ω^2	$1 - \Omega^2$
(1/2, 1/2, 0)	1/2	1/2	(2, 1, 0)	5	–4
(1, 0, 0)	1	0	(3/2, 3/2, 1)	11/2	−9/2
(1, 1/2, 1/2)	3/2	−1/2	(2, 1, 1)	6	−5
(1, 1, 0)	2	−1	(5/2, 1/2, 0)	13/2	−11/2
(3/2, 1/2, 0)	5/2	−3/2	(2, 3/2, 1/2)	13/2	−11/2
(1, 1, 1)	3	−2	(5/2, 1, 1/2)	15/2	−13/2
(3/2, 1, 1/2)	7/2	−5/2	(2, 2, 0)	8	−7
(2, 0, 0)	4	−3	(5/2, 3/2, 0)	17/2	−15/2
(3/2, 3/2, 0)	9/2	−7/2	(2, 3/2, 3/2)	17/2	−15/2
(2, 1/2, 1/2)	9/2	−7/2	(3, 0, 0)	9	−8
(2, 1, 0)	5	−4	(2, 2, 1)	9	−8

Table 2.
Possible structures and invariants for spinor fields (component signs are arbitrary).

Spin is invariant from the point of view of general relativity since it does not depend on the state of motion—the speed and rotation of the reference frame, in which we describe the spinor field.

The spinor properties of fields are unchanged and do not depend on the method and sources of their creation. For each case of motion of a particle in a given field, the spinor properties of the field are determined by the physical nature of the field itself. Spin is a fundamental, unchanging characteristic of the field. Whichever way it is created, it will be with the same back.

Spin in the expression of potential energy describes the system of particle + field, the spinor properties of which are described by the above expressions, regardless of which components are included in the system. The new field of interaction is also represented by one of the above configurations. Whether the spinor field of interaction is a simple sum of spinor fields or another configuration can be found by comparing with experimental data. For each system, a spin configuration should be selected so that the results of the calculation coincide with the experimental data or with the already known characteristics of the system.

The spin of the nucleus and the presence of other electrons with their spins do not determine the spin of the electromagnetic field created by it in atoms. Fine splitting, which directly depends on the spin of the interaction field, does not depend on the spin of the nuclei and electron shell of the atoms. For example, the value of fine splitting of isotope atoms Potassium ^{39}K, ^{41}K, and ^{40}K with spines 3/2, 3/2, and 4; Rubidium ^{87}Rb and ^{85}Rb with spines 3/2 и 5/2; Hydrogen ^{1}H, ^{2}H, ^{3}H with spines 1/2, 1, 1/2 [7].

Also in the solar system, the spin of the gravitational field does not depend on the positions of celestial bodies and their rotation.

Saying that a particle has spin, within the framework of the foregoing, means that the particle has a spinor field, and this property manifests itself at any point in space during any interactions. Spin is a spatial characteristic and is not attributed to any point particle. Therefore, the spin of elementary particles is not determined by their internal structure, but is determined by the spin of the interaction fields created by these particles. The phrase "spin-orbit interaction" in this case means the interaction of the orbital moment with the spinor field.

5. Examples of spinor fields representation

5.1 Electromagnetic wave: photon

The spin of an electromagnetic wave with potential φ is represented as a rotating electric field with a vector potential $\mathbf{A} = \mathbf{\Omega}\varphi$. Given that the magnetic field $\mathbf{B}$ of the wave is defined as the rotor of the vector potential

$$\mathbf{B} = \mathrm{rot}(\varphi\mathbf{\Omega}) = -\left[\mathbf{\Omega} \times \frac{\partial\varphi}{\partial\mathbf{r}}\right] = [\mathbf{\Omega} \times \mathbf{E}], \tag{18}$$

and must satisfy the condition $\mathrm{div}(\varphi\mathbf{\Omega}) = \mathbf{\Omega} \cdot \mathbf{E} = 0,$ get $\mathbf{B}^2 = \mathbf{\Omega}^2\mathbf{E}^2$. So how for an electromagnetic wave $\mathbf{B}^2 = \mathbf{E}^2,\ \ \mathbf{\Omega}^2 = 1$ and the invariant is zero, then the only structure corresponding to these conditions is $\mathbf{\Omega} = (1,\ \ 0,\ \ 0)$ (**Table 2**) with a single projection on the axis of propagation (spin) 1. Thus, the four-dimensional vector of the electromagnetic wave is represented as

$$\mathbf{P} = (\varphi,\ \ \varphi\mathbf{\Omega}) = (\varphi,\ \ \varphi,\ \ 0,\ \ 0) = (\varphi,\ \ \varphi\mathbf{n}),\ \ \mathbf{P}^2 = 0, \tag{19}$$

where n is the unit vector in the direction of wave propagation.

For a monochromatic wave propagating in the direction of the axis $\hat{\mathbf{z}}$ with a potential $\varphi(x,y)$ and wavelength λ, we have

$$\mathbf{P} = (\varepsilon,\ \mathbf{p}) = \varphi(x,y)\exp\left(\frac{\tau - z}{\lambda}i\right)(1,\ \ \ \hat{\mathbf{z}}),\ \ \mathbf{P}^2 = 0,\ \ \ \frac{\partial\varepsilon}{\partial\tau} + \mathrm{div}\mathbf{p} = 0. \tag{20}$$

For transverse waves of other fields, you can select one-component structures with other integers in the form $\mathbf{\Omega} = (m,\ \ 0,\ \ 0),\ \ m = 1, 2, 3, \ldots$ (**Table 2**).

5.2 Stationary magnetic field

$$\begin{gathered}\mathbf{P} = (0,\ \ \varphi\mathbf{\Omega});\quad \mathbf{B} = \mathrm{rot}(\varphi\mathbf{\Omega}) = \left[\mathbf{\Omega} \times \frac{\partial\varphi}{\partial\mathbf{r}}\right] = \\ \left(\Omega_y\frac{\partial\varphi}{\partial z} - \Omega_z\frac{\partial\varphi}{\partial y},\ \ -\ \Omega_x\frac{\partial\varphi}{\partial z} + \Omega_z\frac{\partial\varphi}{\partial x},\ \ \Omega_x\frac{\partial\varphi}{\partial y} - \Omega_y\frac{\partial\varphi}{\partial x}\right) = (0,\ \ 0,\ \ B).\end{gathered} \tag{21}$$

For solenoidal magnetic fields having cylindrical symmetry, we have $\Omega_z = 0$ and $\Omega_x = \Omega_y = 1/2$. The solution for such a system is the vector potential in the form of a structure. $\mathbf{\Omega} = (1/2,\ \ 1/2,\ \ 0), \mathbf{\Omega}^2 = 1/2$ (**Table 2**) with one of the linear functions as an electric field potential

$$\varphi\mathbf{\Omega} = \varphi(1/2,\ \ 1/2,\ \ 0);\ \ \varphi = \varphi_0(x - y);\ \ \mathbf{B} = [\mathbf{\Omega} \times \mathbf{E}] = (0,\ \ 0,\ \ E). \tag{22}$$

5.3 Spinor field with potential $\varphi = \alpha/r$

For the potential in the form of $\varphi = \alpha/r$ we have

$$\mathbf{P} = \left(\frac{\alpha}{r}, \quad \frac{\alpha}{r}\mathbf{\Omega}\right) = \frac{\alpha}{r}(1, \quad \mathbf{\Omega}), \qquad \mathbf{P}^2 = \frac{\alpha^2}{r^2}\left(1 - \mathbf{\Omega}^2\right) \tag{23}$$

The spinor field of this potential can have different structures and invariant values $1 - \mathbf{\Omega}^2$. For Coulomb interaction $\mathbf{\Omega} = (1, \ 1, \ 0), \quad 1 - \mathbf{\Omega}^2 = -1$, and gravity – $\mathbf{\Omega} = (2, \ 2, \ 0), 1 - \mathbf{\Omega}^2 = -7$ (**Table 2**). Such structures and values of the spinor representation provide an accurate description of the spectrum of hydrogen-like atoms, the value of fine splitting, the secular shift of Mercury's perihelion, and the angle of deflection of the light beam by the gravitational field of the Sun.

6. Equations of relativistic and quantum mechanics with spinor fields

Adding spinor fields in the equations of relativistic and quantum mechanics for a particle in an external field (**Table 1**) is not associated with any difficulties, since the spin properties of the field are clearly represented as the vector potential of interaction $\mathbf{A} = \varphi\mathbf{\Omega}$. Equations can be represented as

$$\left(\frac{\partial S}{\partial \tau}\right)^2 - \left(\frac{\partial S}{\partial \mathbf{r}}\right)^2 = \frac{(mc^2 + q\varphi)^2 - (q\varphi)^2\mathbf{\Omega}^2}{c^2}. \tag{24}$$

$$\frac{\partial^2 \psi}{\partial \tau^2} - \frac{\partial^2 \psi}{\partial \mathbf{r}^2} = -\frac{(mc^2 + q\varphi)^2 - (q\varphi)^2\mathbf{\Omega}^2}{\hbar^2 c^2}\psi. \tag{25}$$

Accordingly, the Hamiltonian of the system can be represented as an expression (4)

$$\mathrm{H} = \frac{E^2 - m^2c^4}{2mc^2} = \frac{\mathbf{p}^2}{2m} + q\varphi + \frac{q^2\varphi^2}{2mc^2}\left(1 - \mathbf{\Omega}^2\right). \tag{26}$$

By presenting momentum as $\mathbf{p} = \mathbf{p}' + q\varphi\mathbf{\Omega}/c$, the Hamiltonian can be represented is as

$$\mathrm{H} = \frac{1}{2m}\mathbf{p}'^2 + \frac{q\varphi}{mc}\mathbf{p}' \cdot \mathbf{\Omega} + q\varphi + \frac{q^2\varphi^2}{2mc^2}, \tag{27}$$

and we see that the angular momentum and energy of the particle in the ground state have additional angular momentum and spin state energy. And the expression $q\varphi\mathbf{p}' \cdot \mathbf{\Omega}/mc$is "spin-orbit interaction"—the interaction of a particle with a spinor field.

If other external fields are present, such as a field with a vector potential $\mathbf{A}$, then the equations are represented as

$$\left(\frac{\partial S}{\partial \tau}\right)^2 - \left(\frac{\partial S}{\partial \mathbf{r}}\right)^2 = \frac{(mc^2 + q\varphi)^2 - q^2(\mathbf{A} + \varphi\mathbf{\Omega})^2}{c^2} \tag{28}$$

$$\frac{\partial^2 \Psi}{\partial \tau^2} - \frac{\partial^2 \Psi}{\partial \mathbf{r}^2} = -\frac{(mc^2 + q\varphi)^2 - q^2(\mathbf{A} + \mathbf{\Omega}\varphi)^2}{\hbar^2 c^2}\Psi. \tag{29}$$

The Hamiltonian of the system can be represented as

$$\mathrm{H} = \frac{E^2 - m^2c^4}{2mc^2} = \frac{\mathbf{p}^2}{2m} - q\varphi - \frac{q^2\varphi^2}{2mc^2}\left(1 - \mathbf{\Omega}^2\right) - \frac{q\varphi}{mc^2}\mathbf{\Omega} \cdot \mathbf{A} - \frac{q^2}{2mc^2}\mathbf{A}^2 \tag{30}$$

By presenting momentum as $\mathbf{p} = \mathbf{p}' + q(\varphi\boldsymbol{\Omega} + q\mathbf{A})/c$, the Hamiltonian can be represented is as

$$\mathrm{H} = \frac{1}{2m}\mathbf{p}'^2 + \frac{q}{mc}\mathbf{p}' \cdot (\varphi\boldsymbol{\Omega} + \mathbf{A}) + q\varphi + \frac{q^2\varphi^2}{2mc^2}. \tag{31}$$

7. Some results of solutions of equations of relativistic and quantum mechanics with spinor fields

7.1 Secular shift of Mercury's perihelion

Let us consider the motion of a particle with the mass m in the field of a point body of mass m'. Then the problem reduces to an investigation of the motion of the particle in the centrally symmetric gravitational field with the potential $-Gm'/r$, where G is the gravitational constant, $r_g = Gm'/c^2$.

Choosing the polar coordinates (r, φ) in the plane of motion, we obtain the Hamilton-Jacobi (24) equation in the form

$$\left(\frac{\partial S}{\partial \tau}\right)^2 - \left(\frac{\partial S}{\partial r}\right)^2 - \frac{1}{r^2}\left(\frac{\partial S}{\partial \varphi}\right)^2 = (mc)^2\left(1 - \frac{2r_g}{r} - \left(\boldsymbol{\Omega}^2 - 1\right)\frac{{r_g}^2}{r^2}\right). \tag{32}$$

Let us represent the action S in the form

$$S = -\varepsilon\tau + M\varphi + mcf(r), \tag{33}$$

where ε and M are the constant energy and angular momentum, respectively. As a result, we obtain

$$\left(\frac{\partial f(r)}{\partial r}\right)^2 = \left(\frac{\varepsilon}{mc}\right)^2 - 1 - \frac{2r_g}{r} - \frac{(M/mc)^2 - \left(\boldsymbol{\Omega}^2 - 1\right){r_g}^2}{r^2} \tag{34}$$

and

$$f(r) = \int\sqrt{\left(\frac{\varepsilon}{mc}\right)^2 - 1 + \frac{2r_g}{r} - \frac{(M/mc)^2 - \left(\boldsymbol{\Omega}^2 - 1\right){r_g}^2}{r^2}}dr. \tag{35}$$

We find trajectories from the condition $\partial S/\partial M = 0$, with the use of which we obtain,

$$\varphi = -\int\frac{M/mcr_g}{\sqrt{\left(\frac{\varepsilon}{mc}\right)^2 - 1 + 2\frac{r_g}{r} - \left(\left(M/mcr_g\right)^2 - \left(\boldsymbol{\Omega}^2 - 1\right)^2\right)\frac{r_g^2}{r^2}}}d\frac{r_g}{r}, \tag{36}$$

which results in the solution

$$r = r_g\left(\left(\frac{M}{mcr_g}\right)^2 + 1\right) \times \frac{1}{1 + \sqrt{\left(\frac{\varepsilon}{mc}\right)^2\left(1 + \left(\frac{M}{mcr_g}\right)^2\right) - \left(\frac{M}{mcr_g}\right)^2}\cos\left(\varphi\sqrt{1 - \left(\boldsymbol{\Omega}^2 - 1\right)\left(\frac{mcr_g}{M}\right)^2}\right)}. \tag{37}$$

The secular shift for the gravitational field (23) with $\mathbf{\Omega} = (2,\ 2,\ 0),\ 1 - \mathbf{\Omega}^2 = -7$ is calculated by the formula

$$\Delta\varphi = 2\pi - \frac{2\pi}{\sqrt{1 - \left(\mathbf{\Omega}^2 - 1\right)\left(G\frac{mm'}{cM}\right)^2}} \approx -7\pi\left(G\frac{mm'}{cM}\right)^2. \tag{38}$$

For the Schwarzschild metric, the formula is: $-6\pi(Gmm'/cJ)^2$ [6].

7.2 Photon deflection by the gravitational field of the sun

For a photon whit energy E and equivalent mass$m = E/c^2$, from (20) the Eq. (24) represented in the form

$$\begin{gathered}\left(\frac{\partial S}{\partial \tau}\right)^2 - \left(\frac{\partial S}{\partial \mathbf{r}}\right)^2 = (mc)^2\left(\left(1 - \frac{r_g}{r}\right)^2 - \left(\mathbf{\Omega}_p + \frac{r_g}{r}\mathbf{\Omega}_g\right)^2\right) = \\ (mc)^2\left(-\frac{2r_g}{r} + \mathbf{\Omega}_p \cdot \mathbf{\Omega}_g \frac{2r_g}{r} + \left(1 - {\mathbf{\Omega}_g}^2\right)\frac{{r_g}^2}{r^2}\right).\end{gathered} \tag{39}$$

Choosing the polar coordinates (r, φ) in the plane of motion, we obtain the Hamilton-Jacobi (24) equation in the form

$$\left(\frac{\partial S}{\partial \tau}\right)^2 - \left(\frac{\partial S}{\partial r}\right)^2 - \frac{1}{r^2}\left(\frac{\partial S}{\partial \varphi}\right)^2 = (mc)^2\left(-\frac{2r_g}{r} - \left({\mathbf{\Omega}_g}^2 - 1\right)\frac{{r_g}^2}{r^2}\right) \tag{40}$$

In this case, we used the choice of special directions of the spins of the particle and the field when $\mathbf{\Omega}_p \cdot \mathbf{\Omega}_g = (0, 0, 1) \cdot (2, 2, 0) = 0$(spin-spin interactions are absent).

Let us represent the action S in the form

$$S = -mc\tau + M\varphi + f(r) = -mc\tau + mcR\varphi + f(r), \qquad M = mcR, \tag{41}$$

where M and R are the constant angular momentum and impact parameter, respectively. As a result, we obtain

$$f(r) = mc\int\sqrt{1 + \frac{2r_g}{r} - \frac{R^2 - \left({\mathbf{\Omega}_g}^2 - 1\right){r_g}^2}{r^2}}dr. \tag{42}$$

We find trajectories from the condition $\partial S/\partial M = 0$, with the use of which we obtain,

$$\varphi = -\int\frac{R}{\sqrt{1 + \frac{2r_g}{r} - \frac{R^2 - \left({\mathbf{\Omega}_g}^2 - 1\right){r_g}^2}{r^2}}}d\frac{1}{r}. \tag{43}$$

The photon deflection is calculated by the formula

$$\Delta\varphi = 2\arcsin\frac{1}{\sqrt{1 - \left({\mathbf{\Omega}_g}^2 - 1\right)\left(\frac{r_g}{R}\right)^2}} \approx -2\sqrt{7}\frac{r_g}{R} = -2\sqrt{7}Gm'/c^2R = -5.3Gm'/c^2R. \tag{44}$$

For the Schwarzschild metric, the formula is: $\Delta\varphi = -4Gm'/c^2R$ [6].

7.3 The hydrogen atom problem

For the problem of a hydrogen-like atom without external fields, we have [4].

$$\begin{aligned}&\frac{1}{r^2}\frac{\partial}{\partial r}\left(r^2\frac{\partial\psi}{\partial r}\right)+\frac{1}{r^2\sin\theta}\frac{\partial}{\partial\theta}\left(\sin\theta\frac{\partial\psi}{\partial\theta}\right)+\\&\frac{1}{r^2\sin^2\theta}\frac{\partial^2\psi}{\partial\varphi^2}+\frac{1}{\hbar^2c^2}\left(E^2-\left(mc^2-\frac{Ze^2}{r}\right)^2+\frac{Z^2e^4}{r^2}\boldsymbol{\Omega}^2\right)\psi=0\end{aligned} \tag{45}$$

For the radial part of the solution get

$$\frac{d^2R}{d\rho^2}+\frac{2}{\rho}\frac{dR}{d\rho}-\frac{l(l+1)+(1-\boldsymbol{\Omega}^2)Z^2\alpha^2}{\rho^2}R=-\left(\frac{N}{\rho}-\frac{1}{4}\right)R, \tag{46}$$

Energy levels of the hydrogen atom are (**Figure 11**a)

$$E_{n,l}=mc^2\sqrt{1-\frac{Z^2\alpha^2}{\left(n_r+1/2+\sqrt{(l+1/2)^2-Z^2\alpha^2}\right)^2}}. \tag{47}$$

The fine splitting value

$$E_{2,1}-E_{1,0}\simeq-\frac{Z^2\alpha^2}{12}. \tag{48}$$

The ground states are

$$E_{n,j}=mc^2\sqrt{1-\frac{Z^2\alpha^2}{\left(1/2+\sqrt{(l+1/2)^2-Z^2\alpha^2}\right)^2}},\qquad l+1/2>\sqrt{Z^2\alpha^2+\sqrt{Z\alpha-1/2}}. \tag{49}$$

Ground state with $Z>68$ has a l more than 0, and the greater the atomic number, the discretely it increases in increments of 1.

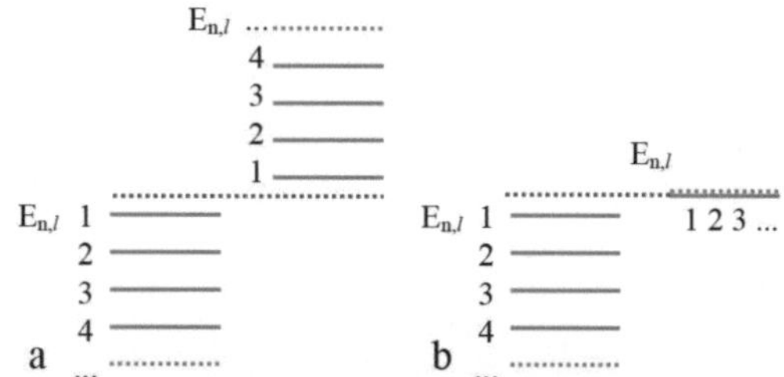

Figure 11.
Fine splitting structures.

From the Dirac equation with $\mathbf{\Omega} = (1,\ 1,\ 0),\ \mathbf{\Omega}^2 = 2$ we get [4].

$$E_{n,j} = mc^2\sqrt{1 - \frac{Z^2\alpha^2}{\left(n_r + \sqrt{(j+1/2)^2 - Z^2\alpha^2}\right)^2}}. \tag{50}$$

The fine splitting value

$$E_{2,3/2} - E_{1,1/2} \simeq -\frac{Z^2\alpha^2}{32}. \tag{51}$$

The ground states are (**Figure 11a**)

$$E_{n,j} = mc^2\sqrt{1 - \frac{Z^2\alpha^2}{(j+1/2)^2 - Z^2\alpha^2}}, \qquad j + 1/2 > \sqrt{2}Z\alpha. \tag{52}$$

Ground state with $Z > 1/\sqrt{2}\alpha \simeq 97$ has a j more than 1/2, and the greater the atomic number, the discretely it increases in increments of 1.

As we can see, the magnitude of the spin-orbit interaction is obtained with a reverse sign and is twice as large as the relativistic splitting of the electron energy levels, which leads to a shift and a change in the order of the splitting levels (**Figure 11a**).

Dirac's solution prompt that if 1/2 is added to the orbital momentum $l \Rightarrow l + 1/2 = j$ and 1 to $n_r \Rightarrow n_r - 1/2$ in solutions without spatial spin (47), then the results of the solutions will be the same. That is, it should be taken into account that in the expression of the orbital moment and energy there are initially present (26–27), independent of the state and fields of the system, a spatial spin with the corresponding value $j = l + 1/2$and energy $n = n_r + l + 1 \Rightarrow n + j + 1/2$.

7.4 Hydrogen atom in a magnetic field

For the hydrogen atom problem in a constant homogeneous magnetic field **B** we have (30)

$$\frac{q^2}{2m^2c^4}\left(\frac{Q}{r}\mathbf{\Omega} - \frac{1}{2}[\mathbf{B}\times\mathbf{r}]\right)^2 = \alpha^2\left(2Z^2\frac{\lambda\!\!\!^-{}^2}{r^2} + \frac{1}{2}\left(\frac{r}{r_h}\right)^2\sin^2\theta + Z\frac{\lambda\!\!\!^-}{r_h}\sin\theta\right), \tag{53}$$

where

$$\lambda\!\!\!^- = \frac{\hbar}{mc}, \qquad \alpha = \frac{e^2}{\hbar c}, \qquad r_h = \frac{c}{\omega}; \qquad \omega = \frac{eB}{2mc}. \tag{54}$$

As we can see, the splitting of levels in a constant magnetic field due to the linear dependence on the field $Z\alpha^2(\lambda\!\!\!^-/r_h)\sin\theta$does not depend on the state of the atom.

7.5 About hyperfine splitting

On hyperfine splitting: why when an electron interacts with nuclear spin, there are only two levels, and a complex structure manifests itself only in the external magnetic field (**Figure 12**).

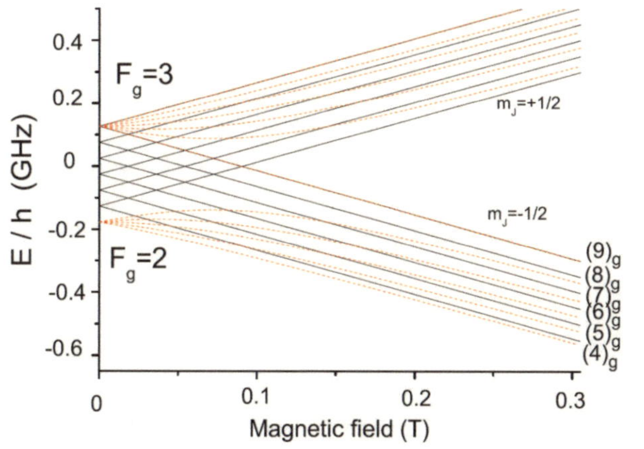

Figure 12.
Theoretical magnetic field dependence of $F_g = 2, 3$ *ground hyperfine levels of* ^{85}*Rb. Red lines: Calculations by the coupled basis theory; black lines: Calculations as given by Eq. (12) (HPB regime). Ground levels for the transitions 4–9 are indicated as (4)g – (9)g [8].*

As has been shown in solutions to the hydrogen-like atom problem for spinor fields, the magnitude of the spin-orbit interaction $\mathbf{L}\cdot\mathbf{S}$is obtained with a reverse sign and is twice as large as the relativistic splitting of the electron energy levels, which leads to a shift and a change in the order of the splitting levels (**Figure 11**a). In the ground state, when the nuclear spin **I** interacts with the spinor field (**Figure 11**b), the spin-spin interaction of the nucleus $\mathbf{I}\cdot\mathbf{S}$, there will be two times less orbital interaction and relativistic splitting (convergence points of black lines in **Figure 12**) will be accurately compensated by the magnitude of the spin-spin interaction of the nucleus (red lines).

8. Representation of the relativistic and quantum mechanics equations in metric spaces

The representation of the equations of relativistic and quantum mechanics in metric spaces means that coordinate transformations have been proposed that bring the equations from **Table 1** to equations with a constant, unit invariant on the right part

$$\begin{aligned}
&\left(\frac{\partial S}{\partial \tau}\right)^2-\left(\frac{\partial S}{\partial \mathbf{r}}\right)^2=I^2 \;\Rightarrow\; g_{ik}\sqrt{-g}\frac{\partial S}{\partial x^i}\frac{\partial S}{\partial x^k}=1,\\
&\frac{\partial^2\psi}{\partial\tau^2}-\frac{\partial^2\psi}{\partial\mathbf{r}^2}=-\frac{I^2}{\hbar^2}\psi \;\Rightarrow\; \frac{1}{\sqrt{-g}}\frac{\partial}{\partial x^i}\left(g_{ik}\sqrt{-g}\frac{\partial\psi}{\partial x^k}\right)=-\psi.
\end{aligned} \tag{55}$$

For clarity, consider the problems with spherical symmetry. For the Hamilton-Jacobi equation in spherical coordinates we have

$$\left(\frac{\partial S}{\partial \tau}\right)^2-\left(\frac{\partial S}{\partial r}\right)^2-\frac{1}{r^2}\left(\frac{\partial S}{\partial \varphi}\right)^2-\frac{1}{r^2\sin^2\theta}\left(\frac{\partial S}{\partial \theta}\right)^2=I(r)^2. \tag{56}$$

and after the coordinates are transformed $\tau, \mathbf{r} \rightarrow \tau', \mathbf{r}'$, they must be represented in metric spaces as

$$\left(\frac{\partial S}{\partial \tau'}\right)^2 - g^2(r')\left(\frac{\partial S}{\partial r'}\right)^2 - \frac{1}{r'^2}\left(\frac{\partial S}{\partial \varphi}\right)^2 - \frac{1}{r'^2 \sin^2\theta}\left(\frac{\partial S}{\partial \theta}\right)^2 = 1. \quad (57)$$

To do this, divide the Eq. (56) by $I(r)^2$

$$\frac{1}{I^2}\left(\frac{\partial S}{\partial \tau}\right)^2 - \frac{1}{I^2}\left(\frac{\partial S}{\partial r}\right)^2 - \frac{1}{I^2 r^2}\left(\frac{\partial S}{\partial \varphi}\right)^2 - \frac{1}{I^2 r^2 \sin^2\theta}\left(\frac{\partial S}{\partial \theta}\right)^2 = 1. \quad (58)$$

and we immediately find an implicit transformation for$r$

$$I(r)r = r'. \quad (59)$$

and

$$\frac{r^2}{r'^2}\left(\frac{\partial S}{\partial \tau}\right)^2 - \frac{r^2}{r'^2}\left(\frac{\partial S}{\partial r}\right)^2 - \frac{1}{r'^2}\left(\frac{\partial S}{\partial \varphi}\right)^2 - \frac{1}{r'^2 \sin^2\theta}\left(\frac{\partial S}{\partial \theta}\right)^2 = 1. \quad (60)$$

It is convenient to choose a new coordinate system associated with a particle, where the velocity is zero $dr'/d\tau' \equiv 0$. Then,

$$d\tau = \frac{\partial \tau}{\partial \tau'}d\tau' + \frac{\partial \tau}{\partial r'}dr' = \left(\frac{\partial \tau}{\partial \tau'} + \frac{\partial \tau}{\partial r'}\frac{dr'}{d\tau'}\right)d\tau' = \frac{\partial \tau}{\partial \tau'}d\tau'.$$

$$\frac{\partial S}{\partial \tau} = \frac{\partial S}{\partial \tau'}\frac{d\tau'}{d\tau} + \frac{\partial S}{\partial r'}\frac{dr'}{d\tau'} = \frac{\partial S}{\partial \tau'}\frac{\partial \tau'}{\partial \tau}. \quad (61)$$

and we get

$$\frac{r^2}{r'^2\left(\frac{\partial \tau}{\partial \tau'}\right)^2}\left(\frac{\partial S}{\partial \tau'}\right)^2 - \frac{r^2}{r'^2\left(\frac{\partial r}{\partial r'}\right)^2}\left(\frac{\partial S}{\partial r'}\right)^2 - \frac{1}{r'^2}\left(\frac{\partial S}{\partial \varphi}\right)^2 - \frac{1}{r'^2 \sin^2(\theta)}\left(\frac{\partial S}{\partial \theta}\right)^2 = 1. \quad (62)$$

$$g(r') = \frac{r}{r'} / \frac{\partial r}{\partial r'} \quad (63)$$

Note that if the $I(r) = const$, then $g(r') = 1$ – only the linear scale changes.

8.1 Field with scalar potential $\varphi(r) = -r_g/r$

For an invariant in the form $I(r) = 1 - r_g/r$ (Coulomb, gravitational field) we find

$$r' = rI(r) = r - r_g, \quad r = r' + r_g, \quad \frac{\partial r}{\partial r'} = 1, \quad (64)$$

$$\tau = \frac{r}{r'}\tau' = \left(1 + \frac{r_g}{r'}\right)\tau', \quad \frac{\partial \tau}{\partial \tau'} = \frac{r}{r'} = 1 + \frac{r_g}{r'}, \quad (65)$$

$$g(r') = \frac{r}{r'} / \frac{\partial r}{\partial r'} = 1 + \frac{r_g}{r'}. \quad (66)$$

From (43), (67), we get

$$\left(\frac{\partial S}{\partial \tau'}\right)^2 - \left(1+\frac{r_g}{r'}\right)^2\left(\frac{\partial S}{\partial r'}\right)^2 - \frac{1}{r'^2}\left(\frac{\partial S}{\partial \varphi}\right)^2 - \frac{1}{r'^2 \sin^2(\theta)}\left(\frac{\partial S}{\partial \theta}\right)^2 = 1. \tag{67}$$

If the field is spinor with invariant $I^2 = (1 - r_g/r)^2 - (\boldsymbol{\Omega} r_g/r)^2$, then from getting

$$r = r_g + \sqrt{r'^2 + \boldsymbol{\Omega}^2 {r_g}^2}, \quad \tau = \frac{r}{r'}\tau' = \left(\frac{r_g}{r'} + \sqrt{1 + \boldsymbol{\Omega}^2 {r_g}^2/r'^2}\right)\tau'. \tag{68}$$

and

$$g^2(r') = \left(1 + \boldsymbol{\Omega}^2 \frac{{r_g}^2}{r'^2}\right)\left(1 + \frac{r_e}{r'} + \frac{\boldsymbol{\Omega}^2 {r_g}^2/r'^2}{1 + \sqrt{1 + \boldsymbol{\Omega}^2 {r_g}^2/r'^2}}\right)^2. \tag{69}$$

8.2 Stationary magnetic field

For a magnetic field with invariant $I^2 = 1 - r^2/{r_B}^2$, where r_B the magnetic event horizon (54), is in cylindrical coordinates we have

$$\left(\frac{\partial S}{\partial \tau}\right)^2 - \left(\frac{\partial S}{\partial r}\right)^2 - \frac{1}{r^2}\left(\frac{\partial S}{\partial \varphi}\right)^2 - \left(\frac{\partial S}{\partial z}\right)^2 = I^2 = 1 - \left(\frac{r}{r_B}\right)^2, \tag{70}$$

and we get (τ and z are transforming similarly)

$$\left(\frac{\partial S}{\partial \tau'}\right)^2 - 2\frac{1-(2r'/r_B)^2}{1+\sqrt{1-(2r'/r_B)^2}}\left(\frac{\partial S}{\partial r'}\right)^2 - \frac{1}{r'^2}\left(\frac{\partial S}{\partial \varphi}\right)^2 - \left(\frac{\partial S}{\partial z'}\right)^2 = 1. \tag{71}$$

$$g^2(r') = 2\frac{1-(2r'/r_B)^2}{1+\sqrt{1-(2r'/r_B)^2}} \tag{72}$$

8.3 Representations of the space-time interval and the wave equation

The results of the metric representations are easily portable to represent the space-time interval (it should be noted that for the space-time interval, the metric tensor is inverse). For example, for a gravitational field with a potential r_g/r', we have (68)

$$\begin{aligned} ds^2 &= I^2\left(d\tau^2 - dr^2 - r^2 d\varphi^2 - r^2 \sin^2(\theta) d\theta^2\right) = \\ &= d\tau'^2 - \frac{dr'^2}{g^2(r')} - r'^2 d\varphi^2 - r'^2 \sin^2(\theta) d\theta^2, \\ ds^2 &= ds'^2 = d\tau'^2 - \frac{1}{(1 + r_g/r')^2} dr'^2 - r'^2 d\varphi^2 - r'^2 \sin^2(\theta) d\theta^2. \end{aligned} \tag{73}$$

and the wave equation in cylindrical and spherical coordinates (τ and z are transforming similarly)

$$\frac{\partial^2\psi}{\partial\tau'^2}-\left(1+\frac{r_g}{\rho'}\right)^2\frac{\partial^2\psi}{\partial\rho'^2}-\frac{1}{\rho'}\left(1+\frac{r_g}{\rho'}\right)\frac{\partial\psi}{\partial\rho'}-\frac{1}{\rho'^2}\frac{\partial^2\psi}{\partial\varphi^2}-\frac{\partial^2\psi}{\partial z'^2}=\psi, \tag{74}$$

$$\frac{\partial^2\psi}{\partial\tau'^2}-\left(1+\frac{r_g}{r'}\right)^2\frac{\partial^2\psi}{\partial r'^2}-\frac{2}{r'}\left(1+\frac{r_g}{r'}\right)\frac{\partial\psi}{\partial r'}-\frac{1}{r'^2}\frac{\partial^2\psi}{\partial\varphi^2}-\frac{1}{r'^2\sin(\theta)}\frac{\partial}{\partial\theta}\left(\sin(\theta)\frac{\partial\Psi}{\partial\theta}\right)=\psi \tag{75}$$

For more complex cases, we will only point out that first, you need to convert the coordinates to bring the expression of the invariant to a diagonal (spherically or cylindrically symmetric) form, and then transform the resulting equation and invariant in the above way.

8.4 Inverse problem

If an equation is given in metric space that has spherical symmetry, it is easy to perform inverse transformations and find the corresponding invariant (field) from (59), (64)

$$g(r')=\frac{r}{r'}\frac{dr'}{dr}; \qquad \frac{dr}{r}=\frac{dr'}{r'g(r')}; \qquad r=e^{\int\frac{dr'}{r'g(r')}}, \qquad I(r)=\frac{r'}{r}. \tag{76}$$

Within the framework of developed approaches, there is an exact correspondence of representations of equations in metric spaces. This is important because it is possible to unambiguously find out which fields correspond to given metric spaces and vice versa.

Naturally, all solutions of equations in metric spaces correspond to solutions of the original equations of relativistic and quantum mechanics having corresponding trajectories of motion and describing the quantum properties of systems.

8.5 Equations in the gravitation field

From the point of view of general relativity, the gravitational field is universal, and all physical processes are described already in metric space. In general, the equations can be represented as follows:

$$g_{ik}\sqrt{-g}\frac{\partial S}{\partial x^i}\frac{\partial S}{\partial x^k}=I^2, \tag{77}$$

$$\frac{1}{\sqrt{-g}}\frac{\partial}{\partial x^i}\left(g_{ik}\sqrt{-g}\frac{\partial\psi}{\partial x^k}\right)=-\frac{I^2}{\hbar^2}\psi. \tag{78}$$

where g_{ik} is the metric tensor determined by a given gravitational field.
After the appropriate conversion $x \rightarrow x'$

$$\begin{gathered}\left(\frac{\partial S}{\partial\tau'}\right)^2-\left(\frac{\partial S}{\partial\mathbf{r}'}\right)^2=I_g{}^2(x')I^2(x(x'))\\ \frac{\partial^2\psi}{\partial\tau'^2}-\frac{\partial^2\psi}{\partial\mathbf{r}'^2}=-\frac{I_g{}^2(x')I^2(x(x'))}{\hbar^2}\psi.\end{gathered} \tag{79}$$

For example, if a hydrogen-like atom is in a constant homogeneous gravitational spinor field $(\varphi_0, \varphi_0 \mathbf{\Omega})$, in the field in which the particle has an invariant $I_g = (1 - 2\varphi_0 - {\varphi_0}^2(\mathbf{\Omega}^2 - 1))$ is considered, we get

$$\left(\frac{\partial S}{\partial \tau'}\right)^2 - \left(\frac{\partial S}{\partial r'}\right)^2 - \frac{1}{r'^2}\left(\frac{\partial S}{\partial \varphi}\right)^2 - \frac{1}{r'^2 \sin^2(\theta)}\left(\frac{\partial S}{\partial \theta}\right)^2 = {I_g}^2{I_e}^2 = (1 - 2\varphi_0 - {\varphi_0}^2({\mathbf{\Omega}_g}^2 - 1))\left(1 - \frac{2r_e}{r(r')} - ({\mathbf{\Omega}_g}^2 - 1)\frac{{r_e}^2}{r^2(r')}\right). \tag{80}$$

A constant homogeneous gravitational field changes the scale of the coordinates by a constant coefficient.

9. Conclusion

Spinor representation of the generalized energy-momentum density 4-vector is proposed, based on the representation of local relativistic rotation by the Lorentz transformation matrix. This representation corresponds to the classical representation of the particle's own rotation, which is described by the diagonal matrix of the moment of inertia.

The spin of the field is invariant from the point of view of general relativity, it does not depend on the state of the system and the sources of its creation.

Spin is a fundamental, unchanging characteristic of the field. Whichever way it is created, it will be with the same back. The spin of particles is not determined by their internal structure, but is determined by the spin of the interaction fields created by these particles.

Spin is a spatial characteristic and is not attributed to any point particle. The phrase "spin-orbit interaction" in this case means the interaction of the orbital moment with the spinor field.

Solutions to the problems of the motion of particles in various external spinor fields are presented. The results of solving these problems show that the developed approach correctly describes the physical properties of the interaction of particles and fields.

The proposed representation of spinor fields applies to the equations of relativistic and quantum mechanics and their representation in metric spaces.

On the other hand, these proposed approaches are still new and need more detailed theoretical and experimental studies and a more developed and rigorous formulation of the mathematical foundations of the new theory.

All aspects and a more complete presentation of the new theory will be presented shortly in the author's book "Relativistic and Quantum Mechanics – with new formulations of principles and theory".

Conflict of interest

The author declares no conflict of interest.

Author details

Vahram Mekhitarian[1,2]

1 Institute for Physical Research of Armenian National Academy of Sciences, Ashtarak, Armenia

2 Russian–Armenian University, Yerevan, Armenia

*Address all correspondence to: vm@ipr.sci.am; vahram.mekhitarian@gmail.com

References

[1] Mekhitarian V. The invariant representation of generalized momentum. Journal of Contemporary Physics. 2012;**47**:249-256. DOI: 10.3103/S1068337212060011

[2] Mekhitarian V. Canonical solutions of variational problems and canonical equations of mechanics. Journal of Contemporary Physics. 2013;**48**:1-11. DOI: 10.3103/S1068337213010015

[3] Mekhitarian V. Equations of relativistic and quantum mechanics and exact solutions of some problems. Journal of Contemporary Physics. 2018;**53**:1-21. DOI: 10.3103/S1068337218020123

[4] Mekhitarian V. Equations of relativistic and quantum mechanics (without spin). In: Bracken P, editor. Quantum Mechanics. London: IntechOpen; 2020. pp. 107-137. DOI: 10.5772/intechopen.93336

[5] Mekhitarian V. The faraday law of induction for an arbitrarily moving charge. Journal of Contemporary Physics. 2016;**51**:108-126. DOI: 10.3103/S1068337216020031

[6] Landau LD, Lifshitz EM. The Classical Theory of Fields. 4th ed. Vol. 2. Oxford: Butterworth-Heinemann; 1980. p. 444

[7] Alkali D. Line Data [Online]. 2022. Available from: https://steck.us/alkalidata/

[8] Sargsyan A, Hakhumyan G, Leroy C, Pashayan-Leroy Y, Papoyan A, Sarkisyan D, et al. Hyperfine Paschen-Back regime in alkali metal atoms: Consistency of two theoretical considerations and experiment. Optical Society of America B. 2014;**31**(5):1046-1053. DOI: 10.1364/JOSAB.31.001046

Chapter 2

Key Outcomes of 5D Relativity

Detlef Hoyer

Abstract

Interstellar travel needs enormous amounts of energy to accelerate payload, structure, and propellant to high speeds. For a travel to a distant star all three have to be huge. Chemical propellant needs more than 90% of the whole start weight, even from an orbit around the Moon. Most of the energy is necessary to increase kinetic energy. Because kinetic energy depends on inertial mass, a reduction of effective energy could reduce the amount of energy needed. As inertial mass and gravitational mass are the same following the equivalence principle of General Relativity and gravity depends on the gravitational constant, a local variation of the gravitational constant might result in variation of the effective mass. A varying scaling between mass/energy and spacetime could also cause a new force as known form the Pioneer anomaly. Kaluza-Klein Theory is General Relativity extended to 5 dimensions (5D). The gravitational constant is substituted by a scalar field making it variable. This scalar field is predicted to change under strong dynamic electromagnetic fields. Deriving the equation of motion from this 5D-metric predicts a fifth force.

Keywords: 5D relativity, 5D electromagnetism, scalar field, fifth force, interstellar travel

1. Introduction

The basic idea of classical Kaluza-Theory [1] is briefly presented. Many following introductory and basic passages and also figures are taken from a conference paper published with the ASCEND 2021 conference [2]. The focus here is on the didactic presentation of the results in the form of 10 key statements, which are listed and explained in different sections. This 5D Relativity theory is named "induced matter theory," because no extra 5D stress-energy-momentum tensor is necessary, but the known 4D stress-energy-momentum tensor is generated or "induced" by curvature of an empty 5D space. The key results are as follows:

i. the 4 off-diagonal elements of the new fifth column $g_{5\alpha}$ represent the 4-vector potential **A** of moving charges

ii. the 5th diagonal element takes the role of the gravity constant

iii. there is an inhomogeneous wave equation for the scalar field with excitation by electromagnetic fields

iv. in the vicinity of a charge the gravitational constant becomes larger

v. the electric field lines of a charge are caused by pressure only, because the positive field energy of the electric field and the negative field energy of the scalar field cancel each other out

vi. a charged black hole in 5D has only one event horizon in contrast to the Reissner-Nordström-solution, which in 4D has two event horizons and a repulsive gravitational field between them

vii. there is a static soliton solution consisting of a cloud of radiation particles of the scalar field (radions, dilatons, scalarons, axions) with kinetic mass around the mass center

viii. a dependence of the elements on the 5th coordinate can generate mass, pressure, momentum, and energy

ix. there is an inhomogeneous wave equation for the scalar field with excitation by variation of the ten 4D gravitiy-potentials with the 5th dimension

x. there is a new rectilinear component of the Lorentz force that causes acceleration that could look like a reduction in inertia.

2. Classical electrodynamics retrospected

Maxwell's eqs. (3D plus time) specify sources and curls for electric field **E** and magnetic field **B** by which the whole fields are determined according to Helmholtz's theorem [3]. As basic mathematical laws they do not reveal which quantity is a cause and which is a consequence. This way the 3rd Maxwell equation

$$\nabla \times \mathbf{E} = -\mu_0 \frac{\partial B}{\partial t} \tag{1}$$

only declares two simultaneously occurring effects always as equal (non-causal). The cause for all electric and magnetic fields are always charges and currents (charges at rest and charges in motion) [4]. The coupling of the fields **E** and **B** by Maxwell's equations means that both appear as a dual entity. So both are parts of one electromagnetic field with six components (3 electric and 3 magnetic) and instead of electricity and magnetism we speak about electromagnetism.

2.1 Electric and magnetic field derived from four scalar potentials

From Maxwell's equations follows that both fields can be derived from potentials. Maxwell's 2nd equation states that there are no sources and only curls for magnetic fields; thus, the magnetic field **B** can be written as rotation of another vector field $\mathbf{A}(\mathbf{r}, t)$ which depends on position and time:

$$\nabla \cdot \mathbf{B} = 0 \Leftrightarrow \mathbf{B} = \nabla \times \mathbf{A} \quad \text{with} \quad \mathbf{A} = (A_1, A_2, A_3) \tag{2}$$

The vector potential **A** can be used in the 3rd Maxwell equation (Eq. (1)) and we get:

$$0 = \nabla \times \mathbf{E} + \mu_0 \frac{\partial}{\partial t} \nabla \times \mathbf{A} = \nabla \times \left(\mathbf{E} + \mu_0 \frac{\partial \mathbf{A}}{\partial t} \right) \tag{3}$$

Because the term in the brackets has a curl of zero, it can be written as a gradient of a scalar potential φ:

$$\mathbf{E} = -\mu_0 \frac{\partial \mathbf{A}}{\partial t} + \nabla \varphi(\mathbf{r}, t) \tag{4}$$

Thus, it is possible to calculate **B** and **E** also from **A** and φ. The electromagnetic potentials introduced in classical theory—formally as helping quantities—appear in other theories as real necessary fields with physical meaning (e.g., Schrödinger-equation, QED, Proca-equation [3]).

2.2 Electromagnetic field tensor: a dual entity derived from 4-dimensional rotation of four vector potential

In special Relativity (4D) the electromagnetic potentials **A** and φ are combined to a 4-vector, setting A_0 or A_4 to φ. The Faraday tensor or field tensor is defined by the 16 components [5]:

$$F_{ik} = \frac{\partial A_k}{\partial x_i} - \frac{\partial A_i}{\partial x_k} \quad \text{with} \quad i, k = 1,2,3,4 \tag{5}$$

These are the partial derivatives of 4-*dimensional* (4D) *rotation* of A_i. They also occur in 3D rotation and in the gradient, that way the Faraday-tensor contains directly electric and magnetic field components. In Cartesian coorinates they represent Eqs. (1) and (2):

$$\begin{aligned} B_x &= \frac{\partial A_3}{\partial x_2} - \frac{\partial A_2}{\partial x_3} & \qquad E_x &= ic\left(\frac{\partial A_4}{\partial x_1} - \frac{\partial A_1}{\partial x_4}\right) \\ B_y &= \frac{\partial A_1}{\partial x_3} - \frac{\partial A_3}{\partial x_1} & \qquad E_y &= ic\left(\frac{\partial A_4}{\partial x_2} - \frac{\partial A_2}{\partial x_4}\right) \\ B_z &= \frac{\partial A_2}{\partial x_1} - \frac{\partial A_1}{\partial x_2} & \qquad E_z &= ic\left(\frac{\partial A_4}{\partial x_3} - \frac{\partial A_3}{\partial x_4}\right) \end{aligned} \tag{6}$$

which were derived from 2nd and 3rd Maxwell's equation. The tensor F_{ik} is the 4D rotation of four vectors A_i. It can be written this way with use of the electromagnetic field quantities:

$$\mathrm{F} = (F_{ik}) = \begin{bmatrix} 0 & B_z & -B_y & -\frac{i}{c}E_x \\ -B_z & 0 & -B_x & -\frac{i}{c}E_y \\ B_y & B_x & 0 & -\frac{i}{c}E_z \\ -\frac{i}{c}E_x & -\frac{i}{c}E_y & -\frac{i}{c}E_z & 0 \end{bmatrix} \tag{7}$$

This tensor is a physical quantity, which unifies electric and magnetic field components into one dual entity: the electromagnetic field tensor F.

2.3 Stress-energy-momentum tensor

The *Maxwell stress tensor* σ (3D) is used in classical electromagnetism to represent the interaction between electromagnetic forces and mechanical momentum ([6], p. 72, 931). Starting with the Lorentz force law $\mathbf{F} = q(\mathbf{E} + \mathbf{v} \times \mathbf{B})$ the force per unit volume is $\mathbf{f} = \rho\ \mathbf{E} + \mathbf{j} \times \mathbf{B}$. Replacing sources ρ and curls $\mathbf{j}$ with Maxwell's equations by functions of the fields **E** and **B**, eliminating the curls by vector calculus identities [7] and writing the result in a more compact way one gets:

$$\sigma_{ij} = \varepsilon_0 E_i E_j + \frac{1}{\mu_0} B_i B_j - \frac{1}{2}\delta_{ij}\left(\varepsilon_0 E^2 + \frac{1}{\mu_0} B^2\right) \quad \text{in} \quad \text{SI units} \tag{8}$$

Which is explicitly in cartesian coordinates

$$\sigma = \begin{pmatrix} \varepsilon_0 E_x E_x + \frac{1}{\mu_0} B_x B_x - \frac{1}{2}(E^2 + H^2) & \varepsilon_0 E_x E_y + \frac{1}{\mu_0} B_x B_y & \varepsilon_0 E_x E_z + \frac{1}{\mu_0} B_x B_z \\ \varepsilon_0 E_y E_x + \frac{1}{\mu_0} B_y B_x & \varepsilon_0 E_y E_y + \frac{1}{\mu_0} B_y B_y - \frac{1}{2}(E^2 + B^2) & \varepsilon_0 E_y E_z + \frac{1}{\mu_0} B_y B_z \\ \varepsilon_0 E_z E_x + \frac{1}{\mu_0} B_z B_x & \varepsilon_0 E_z E_y + \frac{1}{\mu_0} B_z B_y & \varepsilon_0 E_z E_z + \frac{1}{\mu_0} B_z B_z - \frac{1}{2}(E^2 + B^2) \end{pmatrix} \tag{9}$$

This stress tensor can be used to derive the *Lorentz force* per unit volume by the *divergence* of the stress tensor as

$$\mathbf{f} = \text{div}\ \sigma + \varepsilon_0 \mu_0 \frac{\partial \mathbf{S}}{\partial t} \quad \text{with Poynting vector} \quad \mathbf{S} = \frac{1}{\mu_0} \mathbf{E} \times \mathbf{B} \tag{10}$$

With potentials instead of fields (Eq. (6)) and using the definition of the field tensor $F_{\mu\lambda}$ (Eq. (5)) one gets the relativistic electromagnetic stress–energy-momentum tensor $T_{\mu\nu}$ (4D):

$$T_{\mu\nu} = \frac{1}{\mu_0}\left[F_{\mu\lambda}F_{\lambda\nu} + \frac{1}{4}\delta_{\mu\nu}(F_{\kappa\lambda}F_{\kappa\lambda})\right] \tag{11}$$

which is explicitly in matrix form (in SI units):

$$\mathbf{T} = (T_{ik}) = \begin{bmatrix} \frac{1}{2}\left(\varepsilon_0 E^2 + \frac{1}{\mu_0} B^2\right) & S_x/c & S_y/c & S_z/c \\ S_x/c & -\sigma_{11} & -\sigma_{12} & -\sigma_{13} \\ S_y/c & -\sigma_{21} & -\sigma_{22} & -\sigma_{23} \\ S_z/c & -\sigma_{31} & -\sigma_{32} & -\sigma_{33} \end{bmatrix} \tag{12}$$

where $\mathbf{S} = \frac{1}{\mu_0} \mathbf{E} \times \mathbf{B}$ is the Poynting vector, $\sigma_{ij} = \varepsilon_0 E_i E_j + \frac{1}{\mu_0} B_i B_j - \frac{1}{2}\left(\varepsilon_0 E^2 + \frac{1}{\mu_0} B^2\right)\delta_{ij}$ is the Maxwell stress tensor, and the speed of light is $c = \sqrt{\frac{1}{\mu_0 \varepsilon_0}}$. Eq. (10)—the Lorentz force per unit volume - can be rewritten as:

$$\mathbf{f} = \text{div } \mathbf{T} \tag{13}$$

3. Electromagnetism from 5D

Kaluza found that the relativistic electromagnetic stress–energy-momentum tensor **T** from Eqs. (11) and (12) is mathematically structural equal to a part of a 5D curvature tensor. To see this, the electromagnetic vector potential **A**, which had initially 3 spatial components A_x,A_y,A_z and was extended by a fourth (time) component A_t, now is extended again by a fifth component A_l to a 5-vector

$$\mathbf{A} = \left(A_t, A_x, A_y, A_z, A_l\right). \tag{14}$$

The component along the ℓ-axis shall have constant length and may be slanted giving a picture (**Figure 1**) like the hairs of a horses coat or fur, in a pelt or like the tufts of a carpet:

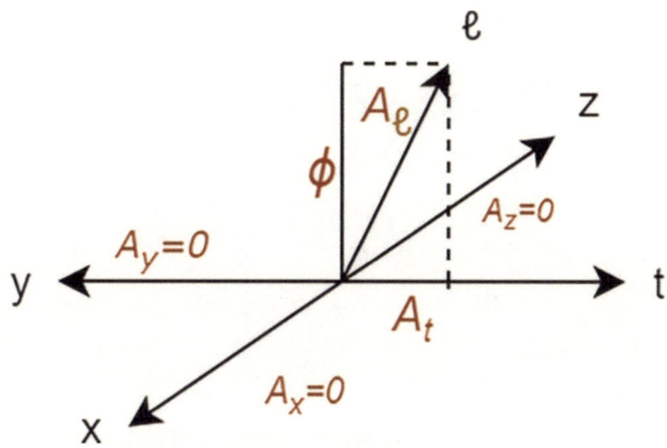

Figure 1.
4-brane (x,y,z,t) with upright 5th component ℓ (axes of a pyramid).

Φ is the component orthogonal to all spacetime components, which are also perpendicular to each other, so the sum

$$A = \sqrt{A_t^2 + A_x^2 + A_y^2 + A_z^2 + \Phi^2} \tag{15}$$

is the length of **A**. To show the capabilities of the Ricci curvature tensor in 5D we take the Minkowski metric and add the 5-vector **A** as fifth row and column.

$$g^{AB} = \begin{pmatrix} 1 & 0 & 0 & 0 & A_t \\ 0 & -1 & 0 & 0 & A_x \\ 0 & 0 & -1 & 0 & A_y \\ 0 & 0 & 0 & -1 & A_z \\ A_t & A_x & A_y & A_z & -\Phi^2 \end{pmatrix} \tag{16}$$

With $A_l = 1$ and $A_t = \varphi = \frac{q}{r}$ the fifth diagonal element becomes

$$\Phi^2 = A_l^2 - A_t^2 = 1 - \varphi^2 = 1 - \frac{q^2}{r^2} \tag{17}$$

and setting $A_x = A_y = A_z = 0$ (no magnetic field) we get with $r = \sqrt{x^2 + y^2 + z^2}$:

$$g^{AB} = \begin{pmatrix} 1 & 0 & 0 & 0 & \frac{\mathbf{q}}{\mathbf{r}} \\ 0 & -1 & 0 & 0 & 0 \\ 0 & 0 & -1 & 0 & 0 \\ 0 & 0 & 0 & -1 & 0 \\ \frac{\mathbf{q}}{\mathbf{r}} & 0 & 0 & 0 & -\left(1 - \frac{\mathbf{q}^2}{\mathbf{r}^2}\right) \end{pmatrix} \tag{18}$$

Applying the 5D Einstein tensor $G^{AB} = R^{AB} - \frac{1}{2}\, g^{AB}\, R$ multiplied by two we get (Maple-instruction: 2*Einstein[~]):

$$2 \cdot G^{AB} = \begin{bmatrix} -\frac{q^2}{r^4} & 0 & 0 & 0 & -\frac{\sqrt{2}q^3}{r^5} \\ 0 & \left(\frac{qx}{r^3}\right)^2 - \frac{q^2}{2r^4} & \frac{qx}{r^3}\frac{qy}{r^3} & \frac{qx}{r^3}\frac{qz}{r^3} & 0 \\ 0 & \frac{qy}{r^3}\frac{qx}{r^3} & \left(\frac{qy}{r^3}\right)^2 - \frac{q^2}{2r^4} & \frac{qy}{r^3}\frac{qz}{r^3} & 0 \\ 0 & \frac{qz}{r^3}\frac{qx}{r^3} & \frac{qz}{r^3}\frac{qy}{r^3} & \left(\frac{qz}{r^3}\right)^2 - \frac{q^2}{2r^4} & 0 \\ \frac{\sqrt{2}q^3}{r^5} & 0 & 0 & 0 & -\frac{2q^2}{r^3} + 1 \end{bmatrix} \tag{19}$$

With: $\vec{e}_r = \frac{x\hat{i}+y\hat{j}+z\hat{k}}{\sqrt{x^2+y^2+z^2}}$, $\vec{E} = \frac{q}{r}\vec{e}_r = \frac{q\left(x\hat{i}+y\hat{j}+z\hat{k}\right)}{\left(x^2+y^2+z^2\right)^{3/2}}$.
an identification of the electric field components in (19) is possible:

$$\begin{aligned} E^2 &= \frac{q^2(x^2+y^2+z)}{\left(x^2+y^2+z^2\right)^3} = \frac{q^2}{r^4} \\ E_x &= \hat{i} \cdot \vec{E} = \frac{qx}{\left(x^2+y^2+z^2\right)^{3/2}} = \frac{qx}{r^3} \\ E_y &= \hat{j} \cdot \vec{E} = \frac{qy}{\left(x^2+y^2+z^2\right)^{3/2}} = \frac{qy}{r^3} \\ E_z &= \hat{k} \cdot \vec{E} = \frac{qz}{\left(x^2+y^2+z^2\right)^{3/2}} = \frac{qz}{r^3} \\ E_x^2 - \frac{1}{2}E^2 &= \frac{q^2x^2}{\left(x^2+y^2+z^2\right)^3} - 1/2\,\frac{q^2}{\left(x^2+y^2+z^2\right)^2} \end{aligned} \tag{20}$$

and it becomes obvious, that the first 4 rows and columns represent the electromagnetic energy-stress-momentum tensor T_{ik} of (11) and (12).

Key result (i): This was only a demonstration which shows that the 5th row and 5th column are related to the electromagnetic 4-vector. Next has to be taken into account, that energy causes a gravitational field.

4. Field of a charged mass at rest in 5D

In the previous section was shown how the 5D Einstein tensor produces the electromagnetic energy-stress-momentum tensor $T_{\mu\nu}$. Because the 5D Einstein tensor already includes the electromagnetic energy-stress-momentum tensor, there is no need for anything besides the Ricci curvature tensor and solutions shall fulfill (symbols with a tilde mark 5D quantities) the 5D equation $\tilde{R}_{AB} = 0$ which infers $\tilde{R} = 0$ and $\tilde{G}_{AB} = 0$.

We start with a 5D line element (with capital S as mark for a 5D quantity) in a form which can be found in ([8], p. 128):

$$dS^2 = \left(\frac{B}{E} - EA^2\right)d(ct)^2 - \frac{1}{B}\ dr^2 - r^2 d\theta^2 - r^2 \sin\theta^2 - 2\ EA\ dt\ dl - E\ dl^2 \tag{21}$$

To be a solution of $R_{AB} = 0$ the functions A, B, and E have to be:

$$E(r) = 1 + \frac{q^2}{2mr}, \quad B(r) = \mathrm{E} - \frac{2m}{r}, \quad A(r) = -\frac{q}{\mathrm{E}\ r} \tag{22}$$

Inserting the solution into the the line element gives:

$$dS^2 = \left(\frac{1 + \frac{q^2}{2Mr} - \frac{2M}{r}}{1 + \frac{q^2}{2Mr}} - \frac{q^2}{1 + \frac{q^2}{2Mr}}\right)d(ct)^2 - \frac{1}{1 + \frac{q^2}{2Mr} - \frac{2M}{r}}dr^2 - r^2 d\theta^2 - r^2 \sin\theta^2$$
$$-2\ \frac{q}{r}\ dt\ dl - \left(1 + \frac{q^2}{2Mr}\right)dl^2 \tag{23}$$

In matrix form this is:

$$\tilde{g}_{AB} = \begin{pmatrix} 1 - \frac{2M}{r} & 0 & 0 & 0 & \frac{q}{r} \\ 0 & \frac{1}{1 - \frac{2M}{r} + \frac{q^2}{2Mr}} & 0 & 0 & 0 \\ 0 & 0 & -r^2 & 0 & 0 \\ 0 & 0 & 0 & -r^2(\sin(\theta))^2 & 0 \\ \frac{q}{r} & 0 & 0 & 0 & -1 - \frac{q^2}{2Mr} \end{pmatrix} \tag{24}$$

Calculating the Ricci curvature with computer algebra software one gets:

$$\tilde{R}_{AB} = \begin{pmatrix} 0 & 0 & 0 & 0 & 0 \\ 0 & 0 & 0 & 0 & 0 \\ 0 & 0 & 0 & 0 & 0 \\ 0 & 0 & 0 & 0 & 0 \\ 0 & 0 & 0 & 0 & 0 \end{pmatrix} \tag{25}$$

The 5D-Monopole metric $\tilde{g}_{AB}$ of Eq. (24) thus is a solution of equation $\tilde{R}_{AB} = 0$.

4.1 Projection of 5D solutions onto 4D space

We start with a common fully covariant 5D line element and use a latin letters to mark 5D components ([8], p. 118)

$$dS^2 = \gamma_{AB}\ dx^A dx^B \tag{26}$$

To get the 4D line element and 4D metric tensor from 5D, the part along the fifth coordinate is split into an orthogonal part and a tangent part in relation to the 4D hyper-surface. A unit vector along the fifth coordinate is (because $\gamma_{55} = \gamma_{51}^2 + \gamma_{52}^2 + \gamma_{53}^2 + \gamma_{54}^2 + \Phi^2$):

$$\Psi^A \equiv \frac{\delta_5^A}{\sqrt{\varepsilon\gamma_{55}}}\gamma_{AB} = \frac{1}{\sqrt{\varepsilon\gamma 55}}(\gamma_{51}, \gamma_{52}, \gamma_{53}, \gamma_{54}, \Phi) \tag{27}$$

This will enable us to split the 5D metric into a part parallel to Ψ^A and a 4D part orthogonal to it. We define a projector:

$$g_{AB} \equiv \gamma_{AB} - \varepsilon\Psi^A\Psi^B \tag{28}$$

where by definition of Ψ follows

$$g_{AB} = \gamma_{AB} - \frac{\gamma_{5A}\gamma_{5B}}{\gamma_{55}} \tag{29}$$

This has only 4 rows and 4 columns:

$$g_{55} = g_{5A} = 0 \tag{30}$$

so the 5D line element becomes the sum of a 4D line element and an extra part:

$$dS^2 = g_{\mu\nu}dx^\mu dx^\nu + \gamma_{55}\left(dx^5 + \frac{\gamma_{5\alpha}}{\gamma_{55}}dx^\alpha\right)^2 \tag{31}$$

$$ds^2 \equiv g_{\mu\nu}dx^\mu dx^\nu \tag{32}$$

Let us introduce a 4-vector and a scalar field

$$A_\mu \equiv \frac{\gamma_{5\alpha}}{\gamma_{55}} \tag{33}$$

$$\Phi^2 = \varepsilon\gamma_{55} \tag{34}$$

With these we get

$$dS^2 = ds^2 + \varepsilon\Phi^2\left(dx^4 + A_\mu dx^\mu\right)^2 \tag{35}$$

This line element rewritten in matrix form is ([8], p. 135):

$$\tilde{g}^{AB} = \begin{bmatrix} g^{\mu\nu} & -A^\mu \\ -A^\nu & -\frac{1}{\Phi^2} + A^\alpha A_\alpha \end{bmatrix},\ \tilde{g}_{AB} = \begin{bmatrix} g_{\mu\nu} - \Phi^2 A_\mu A_\nu & -\Phi^2 A_\mu \\ -\Phi^2 A_\nu & -\Phi^2 \end{bmatrix} \tag{36}$$

With the assumption that the $\tilde{g}_{AB}$ are functions only of x^μ and do not depend on x^5, meaning

$$\frac{\partial}{\partial x^5}\ \tilde{g}_{AB}(x^\mu) = 0 \tag{37}$$

we get the following expression

$$\tilde{G}^{\alpha\beta} \equiv \left[\tilde{R}^{\alpha\beta} - \frac{1}{2}\tilde{g}^{\alpha\beta}\tilde{R}\right] \tag{38}$$

$$= \left[R^{\alpha\beta} - \frac{1}{2}g^{\alpha\beta}R\right] + \frac{1}{2}\Phi^2\left(F^\alpha{}_\lambda F^{\beta\lambda} - \frac{1}{4}g^{\alpha\beta}F^{\mu\nu}F_{\mu\nu}\right) - \frac{1}{\Phi}\left(\Phi^{\alpha;\beta} - g^{\alpha\beta}\Phi^\mu{}_{;\mu}\right) \tag{39}$$

$$= G^{\alpha\beta} + \frac{1}{2}\Phi^2\ T^{\alpha\beta}_{em} + T^{\alpha\beta}_{sc} \tag{40}$$

$$\equiv 0 \tag{41}$$

Key result (ii): here the scalar field plays the role of the gravity constant *via* $\frac{1}{2}\Phi^2$.

For the components with $A = 5$ or $B = 5$ two more equations can be derived: a Maxwell-like equation, where the source depends on the scalar field

$$F^\lambda{}_{\alpha;\lambda} = -\frac{3}{\Phi}\Phi^\lambda F_{\lambda\alpha} \tag{42}$$

$$\Phi^\alpha{}_{;\alpha} = -\frac{1}{4}\Phi^3 F^{\mu\nu}F_{\mu\nu} \tag{43}$$

and a wave-like equation for the scalar field itself—**Key result (iii)**.

5. Projection of a 5D-monopole onto 4D

To get the 4D-metric of the 5D-Monpole solution one has to perform the projection of Eq. (29). Because only the first 4 rows need a projection and components γ_{52}, γ_{53} and γ_{54} are zero, the only projection which has to be done is the one with γ_{51}:

$$g_{11} = \gamma_{11} - \frac{\gamma_{51}^2}{\gamma_{55}} \tag{44}$$

We have $\gamma_{11} = \frac{B}{E} - EA^2$, $\gamma_{51} = -EA$ and $\gamma_{55} = -E$, so

$$g_{11} = \frac{B}{E} - EA^2 - \frac{E^2A^2}{-E} = \frac{B}{E} = 1 - \frac{2m}{rE} = 1 - \frac{2m}{r\left(1 + \frac{q^2}{2mr}\right)} = 1 - \frac{4m^2}{2mr + q^2} \tag{45}$$

The 4D line element is then

$$ds^2 = \frac{B}{E}\ d(ct)^2 - \frac{1}{B}\ dr^2 - r^2 d\theta^2 - r^2 \sin\theta^2 \tag{46}$$

or as matrix

$$g_{\mu\nu} = \begin{pmatrix} -\dfrac{4m^2+2mr+q^2}{2mr+q^2} & 0 & 0 & 0 \\ 0 & \dfrac{2mr}{4m^2-2mr-q^2} & 0 & 0 \\ 0 & 0 & -r^2 & 0 \\ 0 & 0 & 0 & -r^2(\sin(\theta))^2 \end{pmatrix} \tag{47}$$

Now that the 4D metric for the 5D-Monopole is known, one can calculate the 4D-Einstein tensor:

$$G_{\alpha\beta} = \begin{pmatrix} 0 & 0 & 0 & 0 \\ 0 & \dfrac{q^2}{r^2(2mr+q^2)} & 0 & 0 \\ 0 & 0 & \dfrac{q^2(20m^3r+4m^2q^2-4r^2m^2-4mq^2r-q^4)}{4rm(2mr+q^2)^2} & 0 \\ 0 & 0 & 0 & \dfrac{q^2(\sin(\theta))^2(20m^3r+4m^2q^2-4r^2m^2-4mq^2r-q^4)}{4rm(2mr+q^2)^2} \end{pmatrix} \tag{48}$$

which after (38) is equal to $\frac{1}{2}\Phi^2\ T^{\alpha\beta}_{em} + T^{\alpha\beta}_{sc}$ with $\frac{1}{2}\Phi^2 = g$.

As stated, the scalar field plays the role of the gravity constant $g = \frac{1}{2}\left(1 + \frac{q^2}{2mr}\right)$.

Key result (iv): with smaller radius r the gravity constant g increases and the electromagnetic energy generates a stronger gravity field in the vicinity of the charge than from pure 4D theory.

Key result (v): Because G_{00} is the energy component and is zero, this means scalar and electric energy cancel out and there are only pressure terms left to generate force and field lines.

5.1 Comparison with Reissner-Nordström metric

The 4D line element of the Reissner-Nordström metric is known as

$$ds^2 = \left(1 - \frac{2m}{r} + \frac{q^2}{r^2}\right)d(ct)^2 - \frac{1}{1 - \frac{2m}{r} + \frac{q^2}{r^2}}\ dr^2 - r^2 d\theta^2 - r^2 \sin\theta^2 \tag{49}$$

which is as matrix

$$g_{\mu\nu} = \begin{pmatrix} 1 - \dfrac{2m}{r} + \dfrac{q^2}{r^2} & 0 & 0 & 0 \\ 0 & \left(1 - \frac{2m}{r} + \frac{q^2}{r^2}\right)^{-1} & 0 & 0 \\ 0 & 0 & -r^2 & 0 \\ 0 & 0 & 0 & -r^2\ \sin^2(\theta) \end{pmatrix} \tag{50}$$

Both are solution of

$$G_{\mu\nu} = R_{\mu\nu} - \frac{1}{2}R = \frac{q^2}{r^4}\ g_{\mu\nu}(= T_{\mu\nu}) \tag{51}$$

The electromagnetic stress-energy-momentum tensor as matrix:

$$T_{\mu\nu} = \begin{pmatrix} \frac{-2mq^2r + q^4 + q^2r^2}{r^6} & 0 & 0 & 0 \\ 0 & \frac{q^2}{2mr^3 - q^2r^2 - r^4} & 0 & 0 \\ 0 & 0 & -\frac{q^2}{r^2} & 0 \\ 0 & 0 & 0 & -\frac{q^2}{r^2}\sin^2(\theta) \end{pmatrix} \tag{52}$$

Key result (vi): The Reissner-Nordström solution has a negative inverse r term $\frac{2m}{r}$ and a positive inverse r^2 term $\frac{q^2}{r^2}$ which can result in 2 event horizons. The 4D projection of the 5D Monopole can have only one event horizon.

Both differ only in the near field. The expansion of g_{00} of the 4D projection of the 5D Monopole shows a vanishing difference to g_{00} of the Reissner-Nordström solution with increasing r:

$$\mathbf{1} - \frac{\mathbf{2m}}{\mathbf{r}} + \frac{\mathbf{q}^2}{\mathbf{r}^2} - \frac{1}{2}\frac{q^4}{mr^3} + \frac{1}{4}\frac{q^6}{m^2r^4} - \frac{1}{8}\frac{q^8}{m^3r^5} + O(r^{-6}) \tag{53}$$

6. Neutral soliton solution with a static scalar field

Setting the electromagnetic 4-potential components $g_{5\alpha} = 0$ results in an electrically neutral solution. Only g_{55} is nonzero and varies with distance r. With a function

$$A(r) = 1 - \frac{2m}{r} \tag{54}$$

the line-element of the solution is

$$ds^2 = A^a dt^2 - A^{-(a+b)}dr^2 - A^{1-a-b}r^2\left(d\theta^2 + \sin\theta^2\right) - A^b d\ell^2 \tag{55}$$

which is in matrix form

$$\begin{pmatrix} \left(1-\frac{2m}{r}\right)^a & 0 & 0 & 0 & 0 \\ 0 & -\left(1-\frac{2m}{r}\right)^{-a-b} & 0 & 0 & 0 \\ 0 & 0 & -r^2\left(1-\frac{2m}{r}\right)^{1-a-b} & 0 & 0 \\ 0 & 0 & 0 & -r^2\sin(\theta)^2\left(1-\frac{2m}{r}\right)^{1-a-b} & 0 \\ 0 & 0 & 0 & 0 & \left(1-\frac{2m}{r}\right)^b \end{pmatrix} \tag{56}$$

Calculating the Ricci tensor for this metric results in

$$\begin{pmatrix} 0 & 0 & 0 & 0 & 0 \\ 0 & \dfrac{2m^2(a^2+2ab+b^2-1)}{r^2(-r+2m)^2} & 0 & 0 & 0 \\ 0 & 0 & 0 & 0 & 0 \\ 0 & 0 & 0 & 0 & 0 \\ 0 & 0 & 0 & 0 & 0 \end{pmatrix} \tag{57}$$

and requests the condition:

$$a^2 + ab + b^2 - 1 = 0 \tag{58}$$

Since $a \neq b$ in general, the metric (52) is bivalent in the sense that it has gravitational and scalar contributions to the energy. Calculating the total energy of a soliton gives (a + b/2) M [7] where M is the mass seen from infinity.

Key result (vii): since there are no electromagnetic sources in the metric (56) and all of its associated matter outside of the central source comes from the last term, the logical consequence is that the cloud of radiation around a soliton is not photons but scalerons, or quanta of the scalar field (radions, dilatons, axions).

6.1 Neutral soliton solution with a dynamic scalar field

There is also a dynamic (isotropic) solution of a soliton, where the scalar field quanta vanish to infinity whith an analog of a Hubble-constant leading to a Black Hole without a scalar field:

$$\begin{aligned} ds^2 = {} & \left(\frac{ar-1}{ar+1}\right)^{\frac{4}{\sqrt{3}}} dt^2 - \left(\frac{a^2r^2-1}{a^2r^2}\right)\left(\frac{ar-1}{ar+1}\right)^{\frac{2}{\sqrt{3}}} (1+Ht)d\sigma^2 \\ & - \left(\frac{a^2r^2-1}{a^2r^2}\right)\left(\frac{ar-1}{ar+1}\right)^{\frac{2}{\sqrt{3}}} (1+Ht)^{-1}\, d\ell^2 \end{aligned} \tag{59}$$

7. Induced matter

In the preceding section, there was no explicit dependence on ℓ. Now, we allow all potentials g_{AB} to depend on ℓ and again work with electric neutral matter (see matrix form in Eq. (36) with condition Eq. (37)):

$$g_{\alpha\beta} = g_{\alpha\beta}(x^A) \qquad g_{5\alpha} = 0 \qquad g_{55} = \Phi(x^A)^2 \tag{60}$$

$$dS^2 = g_{AB}\, dx^A dx^B \tag{61}$$

Next, we can split the 5D Ricci tensor in its 4D analog and terms belonging to the 5th dimension, which can be taken as terms of a stress-energy-momentum tensor. Derivations of the 5th dimension with respect to ℓ are denoted with a star:

$$\tilde{R}_{\alpha\beta} = R_{\alpha\beta} + \left(\Gamma^4_{\alpha\beta}\right)_{,4} - \left(\Gamma^4_{\alpha 4}\right)_{,\beta} + \Gamma^\lambda_{\alpha\beta}\Gamma^4_{\lambda 4} + \Gamma^4_{\alpha\beta}\Gamma^D_{4D} - \Gamma^4_{\alpha\lambda}\Gamma^\lambda_{\beta 4} - \Gamma^D_{\alpha 4}\Gamma^4_{\beta D}$$

$$= R_{\alpha\beta} - \frac{\Phi_{\alpha;\beta}}{\Phi} + \frac{1}{2\Phi^2}\left(\frac{\overset{*}{\Phi}\overset{*}{g}_{\alpha\beta}}{\Phi} - \underset{\alpha\beta}{\overset{**}{g}} + g^{\mu\nu}\ \underset{\alpha\lambda}{\overset{*}{g}}\ \underset{\beta\mu}{\overset{*}{g}} - \frac{g^{\mu\nu}\ \underset{\mu\nu}{\overset{*}{g}}\ \underset{\alpha\beta}{\overset{*}{g}}}{2}\right) \tag{62}$$

$$= R_{\alpha\beta} - T_{\alpha\beta} + \frac{1}{2}\ g_{\alpha\beta}\ T \equiv 0$$

Key result (viii): the dependence of the elements on the coordinate can generate mass, pressure, momentum and energy (tensor $\mathbf{T}_{\alpha\beta}$).

The fifth diagonal element of the 5D-Ricci tensor is also required to be zero:

$$\tilde{R}_{55} = \Phi\Box\Phi - \frac{*g^{*\lambda\beta}\ \ *g_{\lambda\beta}}{2} - \frac{g^{\overset{**}{\lambda\beta}}\ \ *g_{\lambda\beta}}{2} - \frac{*\Phi\ g^{\lambda\beta}\ \ *g_{\lambda\beta}}{2\Phi} - \frac{g^{\mu\beta}\ g^{\lambda\sigma}\ \ *g_{\lambda\beta}\ \ *g_{\mu\sigma}}{4} \equiv 0 \tag{63}$$

Key result (ix): and yields a wave equation for the gravitational constant.

8. Geodesic motion in 5D and extension of Lorentz force

In the preceding section, a new energy-momentum-tensor $T^{\alpha\beta}_{sc}$ occurred. Its divergence should yield in force density (force per unit inertial mass). For that reason, the equation of motion is interesting. Using the Lagrangian approach minimizing the distance between two points in 5D, the equation of geodesic motion becomes ([7], p. 154):

$$\frac{du^\mu}{ds} + \Gamma^\mu_{\beta\gamma}u^\beta u^\gamma + \left(g^{\mu\alpha} - \frac{1}{2}\frac{dx^\mu}{ds}\frac{dx^\alpha}{ds}\right)\frac{d\ell}{ds}\frac{dx^\beta}{ds}\frac{\partial g_{\alpha\beta}}{\partial\ell} = 0 \tag{64}$$

The first two terms describe geodesic motion in 4D, so we split the equation and define a force density:

$$\frac{du^\mu}{ds} + \Gamma^\mu_{\beta\gamma}u^\beta u^\gamma = f^\mu \tag{65}$$

$$f^\mu \equiv \left(-g^{\mu\alpha} + \frac{1}{2}\frac{dx^\mu}{ds}\frac{dx^\alpha}{ds}\right)\frac{d\ell}{ds}\frac{dx^\beta}{ds}\frac{\partial g_{\alpha\beta}}{\partial\ell} = 0 \tag{66}$$

The new force density is nonzero if the 4D metric depends on the fifth coordinate $\left(\frac{\partial g_{\alpha\beta}}{\partial\ell}\right)$ and there is motion in the fifth dimension $\left(\frac{d\ell}{ds}\right)$. A force F^μ can be split into a normal component N^μ in relation to the 4D velocity and a tangent or parallel component P^μ in relation to the 4D velocity, thus $F^\mu = N^\mu + P^\mu$ giving force densities:

$$n^\mu = \left(-g^{\mu\alpha} + u^\mu u^\alpha\right)u^\beta\frac{\partial g_{\alpha\beta}}{\partial\ell} \tag{67}$$

$$p^\mu = -\frac{1}{2}u^\mu\left(u^\alpha u^\beta\frac{\partial g_{\alpha\beta}}{\partial\ell}\right)\frac{d\ell}{ds} \tag{68}$$

Key result (x): the force **P** *parallel* to 4D velocity **u** is new and occurs only if there is fifth dimension dependency ([7], p. 157). In the preceding section, we had condition (37) wherefore there a fifth force does not occur.

9. Conclusions

From five dimensional metrics with unrestricted dependence from all five coordinates resolving the equation $R_{AB} = 0$ many inductively gained physical laws can be deduced: Newton's gravity law, Coulomb law, Faraday induction law, Ampère's circuital law, Maxwell Equations, Einstein's Gravity law, Einstein-Maxwell equation. Because of further degrees of freedom a scalar wave equation for the gravity constant is gained, yielding in an additional energy and an additional force. For 4D assemblies a reduction of inertia is predicted. The gravity constant corresponds to a scalar field which is part of the 5D metric as the 15th potential.

This 15th component of the metric varies in the vincinity of a charge and seems to behave like the potentials of the electromagnetic 4-vector-potential which is able detach from the sources, giving reason to the assumption that changes of the gravity constant could emerge into the surrounding spacetime.

Aiming at the reduction of inertia and an accelerating fifth force, for interstellar travel or asteroid deflection missions a change from analytical to numerical calculation methods will be necessary.

Recent experiments investigate the possible effects of a fifth force with neutron beam scattering on silicon crystals [9]. Neutrons have internally two negative charges and a double positive charge in the smallest space and in the near field the scalar potential causes the greatest deviations.

Acknowledgements

The author would like to thank the American Institute of Aeronautics and Astronautics (AIAA) for kindly allowing parts of this chapter to be adapted from a conference paper published in the ASCEND 2021 conference. Since most of the content presented is introductory and basic, many passages and figures are taken from that paper. The focus here is on the didactic presentation of the results in the form of 10 key statements.

Nomenclature

a	acceleration
A	electromagnetic vector potential
A_μ	component of electromagnetic 4-vector
B	magnetic field
β	rapidity $\frac{v}{c}$
c	velocity of light, displacement of a field source
ds	line element, 4D
dS	line element, 5D
d*t*	time step multiplied with c (velocity of light)
E	electric field

ε	sign + or – (factor + 1 or − 1)
ε_0	dielectric constant
f	force density (force per unit inertial mass)
F	force, electromagnetic Field tensor, Farady tensor
$F_{\mu\nu}$	component of the Farady tensor
φ	electric potential
Φ	component along the 5th coordinate
g	metric tensor, 4D
$g_{\mu\nu}$	component of 4D metric tensor
g_{AB}	component of 5D metric tensor
γ	gravity constant, index for 4D, Lorentz factor $1/\sqrt{1-\beta^2}$, 5D metric tensor
Γ^i_{kl}	Christoffel symbol of the second kind
$\hat{i}, \hat{j}, \hat{k}$	Cartesian unit basis vector (Maple notation)
j	current density
ℓ	fifth coordinate (besides x, y, z, ct)
m	mass
μ_0	magnetic permeability
q	charge
r	radial distance
R	curvature scalar
$R_{\mu\nu}$	Ricci curvature tensor, 4D
R_{AB}	Ricci curvature tensor, 5D
$R^{\sigma}_{\mu\nu\kappa}$	Riemann curvature tensor
ρ	charge density
S	Poyting vector
σ	Maxwell stress tensor
$T_{\mu\nu}$	electromagnetic stress energy tensor
v	velocity
x,y,z	Cartesian component indexes (of a force)
x_i	covariant coordinates
x^i	contravariant coordinates

Author details

Detlef Hoyer†
Technical University of Hamburg, Germany

*Address all correspondence to: d.hoyer@hamburg.de

†Alumnus.

References

[1] Kaluza T. Zum Unitätsproblem der Physik, Sitzungsberichte der Preussischen Akademie der Wissenschaft. Berlin: Phys.-Math. Klasse; 1921. p. 966

[2] Hoyer D. Fast interstellar space travel by reduced mass via 5D Electromagnetism. [Internet]. ASCEND 2021. American Institute of Aeronautics and Astronautics; 2021. DOI: 10.2514/6.2021-4002

[3] Lehner G. Elektromagnetische Feldtheorie für Ingenieure und Physiker. Berlin, Heidelberg: Springer; 1990

[4] Jefimenko O. Causality, Electromagnetic Induction and Gravitation. Star City: Electric Scientific Company; 1992. p. 16

[5] Jackson JD. Klassische Elektrodynamik. 2. Auflage ed. Berlin New York: de Gruyter; 1983

[6] Simonyi K. Introduction. In: Theoretische Elektrotechnik. Berlin: VEB Deutscher Verlag der Wissenschaften; 1989. p. 929

[7] Spiegel MR. Ch. VII. In: Vectoranalysis. Düsseldorf: McGraw-Hill; 1977. p. 17

[8] Wesson P, Overduin J. Space-Time-Matter. Singapore: World Scientific Publishing Co.; 2019

[9] Heacock B. Pendellösung interferometry probes the neutron charge radius, lattice dynamics, and fifth forces. Science. 2021;**373**:1239

Chapter 3

Dissipative Quantum System and Energy Balance

Jishad Kumar

Abstract

We discuss how various parts of a quantum many-body system exchange energies at thermal equilibrium. To show this, we assume a quantum system is coupled to a many-body environment (at thermal equilibrium with a bigger environment) consisting of a large number of independent and non-interacting quantum harmonic oscillators above a stable ground state. Once the coupling to a large environment is switched on, the system dissipates its energy continuously to the environment until it reaches equilibrium with the latter. We use the Quantum Langevin equation to show such energy exchange at equilibrium. We conclude that different parts of a physical system can exchange energies even at absolute zero temperature.

Keywords: open systems, quantum dissipation, fluctuations, instantaneous power, the charged oscillator in a magnetic field

1. Introduction

Isolating a quantum system from its environment is not possible since the coupling energy plays a pivotal role in the low-temperature properties of the system. Moreover, a complete understanding or control of the huge environment is also not feasible. How one will study the properties of the quantum system? A working method is to partition the whole system into two different parts, viz., the *system* and the *environment or heat bath*, and eliminate the bath degrees of freedom after carefully considering the effects of the heat bath on the system parameters. This makes the system essentially an *open* one and the study of open systems has been very crucial in many applications of quantum mechanics [1]. There are consequences due to the establishment of a coupling between the system and the heat bath. Firstly, there may be an irreversible transfer of energy from the system to the environment (*dissipation*), next there is *Brownian motion* - fluctuations in the system's degrees of freedom due to the noisy force exerted on the system by the environment. There is *decoherence*, a purely quantum mechanical phenomenon where the system-bath coupling destroys the coherent superposition of states. The first two processes have classical counterparts which have been extensively studied by many authors in the literature. A more detailed understanding shows that the fluctuating force from the environment induces decoherence and damping in the system properties.

Figure 1.
(a) Schematic representation of Brownian motion. The jittering motion of a large mass immersed in the medium containing a large number of particles/molecules is the phenomenon of Brownian motion. The random kicks the mass received from its environmental particles make the motion stochastic. (b) The famous system-plus-bath arrangement. The test system with one or few degrees of freedom is assumed to be in contact with its environment containing a large number of independent and identical harmonic oscillators. The individual masses of the oscillators in the environment $m_1, m_2, m_3 \dots m_N$ are assumed to be smaller than the mass M of the test system.

The well-known model of the system-plus-bath approach to open quantum systems stems from classical physics. The underlying phenomenon is the classical Brownian motion where a test system undergoes random motion due to the "kicks" received by the former from the surrounding particles when immersed in a suitable medium (**Figure 1**). In the classical setting, dissipation is introduced in the system "by hand" by inserting a time-dependent damping term into the equation of motion. However, this naive approach never works in the quantum domain where everything is governed by certain principles like Heisenberg's uncertainty relation. Moreover, the physical quantities are operators in quantum mechanics and these operators must satisfy certain commutation relations. The damping term inserted in the equation of motion violates the uncertainty principle. The role of fluctuating/random forces is crucial in order to preserve the canonical structure. The knowledge of the details of the processes including dissipation in the system may not be explicitly known so sometimes the dissipation mechanism is globally described by friction, resistance, viscosity, etc. These parameters are introduced in order to compensate for the information loss due to dissipation. In this microscopic system-bath model, *friction* comes about by the irreversible transfer of energy from the system to its environment. The environment is modeled such that no energy, which is transferred, may come back to the system within any physically relevant time periods. That means the so-called Poincaré recurrence time is infinity. A really necessary condition for the full Hamiltonian is that, under certain conditions, the known classical results must be recovered. Quantum dynamics at arbitrarily low temperatures and (or) with strong damping can be studied within the system-plus-bath approach, regardless of whether the bath is ohmic, sub-ohmic, or super-ohmic. The key thermodynamical quantity of a quantum dissipative system is the reduced density operator $\rho(t) = tr_B\rho_T(t)$, i.e., the partial trace of the total system plus bath density operator ρ_T over the bath degrees of freedom. Here t denotes time. Quantum dissipation theory describes not only the evolution of $\rho(t)$ but also the equilibrium behavior of the reduced system as $\rho(t \to \infty) = \rho_{eq}(T) \propto tr_B e^{-\mathcal{H}_T/k_B T}$, where T is the temperature, k_B is the Boltzmann constant and $\mathcal{H}_T$ is the total Hamiltonian. The latter property is also referred as the detailed balance relation of the quantum dissipation theory [2]. The description of the dissipative system is recovered by the reduced density matrix obtained by eliminating the bath degrees of freedom which imparts damping and fluctuations.

The system plus reservoir (bath) approach to open quantum systems, was originally introduced by many authors [3–9] and popularized later in the literature by many others [10–12]. The idea here is to couple a system with a finite degree of freedom (system under study) with a reservoir consisting of an infinite number of independent and non-interacting harmonic oscillators. This model has been discussed in the literature both for harmonic systems [7, 8, 13, 14] and anharmonic systems [15]. Once the coupling is established the reservoir imparts fluctuations in the system coordinates which thereby causes the system under observation to lose energy rapidly (and is irrevocable) to the bath. Because of this fluctuating or random force, the system undergoes Brownian motion. The reservoir is commonly known as a heat bath because the system dissipates its energy continuously and the former distributes this dissipated energy it its various energy-infusing modes. The relevant variables of the heat bath are averaged out later from the larger Hilbert space of the full system-plus-bath setup, to obtain an effective description of the test system alone. The projected dynamics of the test quantum system belonging to the truncated Hilbert space appear dissipative due to the bath-induced decoherence effects. Usually, either a formal path integral approach in the Schrödinger picture [16] or the quantum Langevin equation in the Heisenberg picture [12] is used to eliminate the heat bath degrees of freedom.

2. Theoretical framework

The system-plus-bath model for dissipative quantum systems is described as follows. A quantum system of a finite degree of freedom is coupled to a heat bath consisting of independent and non-interacting harmonic excitations above a stable ground state. The interaction between the quantum system and an individual oscillator of the heat bath is inversely proportional to the total volume $\mathcal{V}$ of the bath, thereby ensuring that the individual coupling is a linear function of bath coordinates. This nice property further allows one to eliminate the bath degrees of freedom easily. Because the number of oscillators in the bath is very large, the weak perturbation of any individual bath oscillator on the quantum system does not necessarily mean that the coupling of the system and the bath is weak. This model, even if the individual oscillators of the bath couple weakly to the system, allows the inclusion of strong damping also [11].

We write the total Hamiltonian for the "full" system as

$$H = H_S + H_B + H_{SB}, \tag{1}$$

where the system Hamiltonian is given by

$$H_S = \frac{p^2}{2M} + V(q), \tag{2}$$

where M is the mass of the quantum system which is moving in a potential $V(q)$, with q being the coordinate of the system. The heat bath Hamiltonian is written as the sum of N non-interacting oscillators

$$H_B = \sum_{j=1}^{N} \left(\frac{p_j^2}{2m_j} + \frac{m_j}{2} \omega_j^2 x_j^2 \right). \tag{3}$$

The possibility of revival of the initial state after a course of time, since one can pass on to the normal coordinates with the heat bath consisting of harmonic oscillators and $V(q)$ as a harmonic potential, can be overcome with heat bath having sufficiently many oscillators so that the Poincaré recurrence time becomes infinity [17]. The third contribution, the interaction Hamiltonian can be written as

$$H_{SB} = -q \sum_{j=1}^{N} C_j x_j + q^2 \sum_{j=1}^{N} \frac{C_j^2}{2m_j \omega_j^2}, \tag{4}$$

which is bilinear in the system and bath coordinates. The last term (which depends on the coupling constants C_j and only contains an operator in the system Hilbert space) in the interaction Hamiltonian is included as a counter term to ensure that the dissipation is homogeneous in all spaces. Since the coupling is via the position variables, if this term is not included, then the coupling becomes different wherever the quantum particle is located. Or in other words, the model is not translationally invariant and that will result in an unphysical renormalization of the potential. Therefore, one must understand that the counter term in the interaction Hamiltonian is included to make sure of the fact that dissipation has been introduced solely by the coupling to the reservoir not by a renormalization of $V(q)$. If it was not included, then the minimum of the potential surface of the global system for a given q is at $x_j = C_j q / m_j \omega_j^2$ for all j . This result in an 'effective' potential renormalized by the coupling which is given by

$$V_{eff}(q) = V(q) - \sum_{j=1}^{N} \frac{C_j^2 q^2}{2m_j \omega_j^2}. \tag{5}$$

This becomes clear if we consider the minimum of the Hamiltonian with respect to the system and environment coordinates. From the requirement

$$\frac{\partial H}{\partial x_j} = m_j \omega_j^2 x_j - C_j q = 0, \tag{6}$$

we obtain

$$x_j = \frac{C_j}{m_j \omega_j^2} q. \tag{7}$$

Using this result, we determine the minimum of the Hamiltonian with respect to the system coordinate and is given by

$$\frac{\partial H}{\partial q} = \frac{\partial V}{\partial q} - \sum_{j=1}^{N} C_j x_j + q \sum_{j=1}^{N} \frac{C_j^2}{m_j \omega_j^2} = \frac{\partial V}{\partial q}. \tag{8}$$

The second term in Eq. (4) thus ensures that this minimum is determined by the potential $V(q)$ only. Before we start writing equations of motion, for simplicity, we write the full Hamiltonian as

$$\mathcal{H} = \frac{p^2}{2M} + V(q) + \sum_{j=1}^{N} \left\{ \frac{p_j^2}{2m_j} + \frac{1}{2} m_j \omega_j^2 \left(x_j - \frac{C_j q}{m_j \omega_j^2} \right)^2 \right\}. \tag{9}$$

2.1 Quantum mechanical derivation

We need to look at the reduced equation of motion for the system coordinate. In this section, we derive the quantum Langevin equation for our test system coordinate under the influence of the heat bath induced fluctuation effects. In the quantum domain, all the parameters are quantum variables and are operators. In the literature, Magalinskiĭ [14] showed that the elimination of the environmental degrees of freedom leads to a damped equation of motion for the system coordinate. The time evolution of an operator A, in the Heisenberg picture, is given by

$$\frac{dA}{dt} = \frac{i}{\hbar}[H, A]. \tag{10}$$

From Eq. (9) we obtain the equations of motion for the bath degrees of freedom

$$\dot{p}_j = -m_j \omega_j^2 x_j + C_j q \tag{11}$$

$$\dot{x}_j = \frac{p_j}{m_j}, \tag{12}$$

and, similarly, the equations of motion for the system degree of freedom are given by

$$\dot{p} = -\frac{\partial V}{\partial q} + \sum_{j=1}^{N} C_j x_j - q \sum_{j=1}^{N} \frac{C_j^2}{m_j \omega_j^2}, \tag{13}$$

$$\dot{q} = \frac{p}{M}. \tag{14}$$

We treat the system coordinate $q(t)$ as if it were a given function of time, we then solve the environmental equations of motion and it turns out to be an ordinary second order linear inhomogeneous differential equation with the solution of the form

$$x_j(t) = x_j(0)\cos(\omega_j t) + \frac{p_j(0)}{m_j \omega_j}\sin(\omega_j t) + \frac{C_j}{m_j \omega_j}\int_0^t ds \sin(\omega_j (t-s)) q(s). \tag{15}$$

Inserting Eq. (15) into Eq. (13), we obtain an effective equation of motion for the system coordinate

$$M\ddot{q} - \int_0^t ds \sum_{j=1}^{N} \frac{C_j^2}{m_j\omega_j} \sin\left(\omega_j(t-s)\right)q(s) + \frac{\partial V}{\partial q} + q\sum_{j=1}^{N} \frac{C_j^2}{m_j\omega_j^2} \tag{16}$$
$$= \sum_{j=1}^{N} C_j \left[x_j(0)\cos\left(\omega_j t\right) + \frac{p_j(0)}{m_j\omega_j}\sin\left(\omega_j t\right)\right].$$

This is further simplified by partially integrating the second term of the LHS and that yields

$$M\ddot{q} + M\int_0^t ds\gamma(t-s)\dot{q}(s) + \frac{\partial V}{\partial q} = f(t), \tag{17}$$

where the damping kernel is given by

$$\gamma(t) = \frac{1}{M}\sum_{j=1}^{N} \frac{C_j^2}{m_j\omega_j^2}\cos\left(\omega_j t\right). \tag{18}$$

It can also be expressed alternatively as

$$\gamma(t-t') = \Theta(t-t')\frac{1}{M}\sum_{j=1}^{N} \frac{C_j^2}{m_j\omega_j^2}\cos\left[\omega_j(t-t')\right]. \tag{19}$$

The operator-valued random force in Eq. (17) takes the form

$$f(t) = \sum_{j=1}^{N} C_j \left[\left(x_j(0) - \frac{C_j}{m_j\omega_j^2}q(0)\right)\cos\left(\omega_j t\right) + \frac{p_j(0)}{m_j\omega_j}\sin\left(\omega_j t\right)\right]. \tag{20}$$

The statistical average of this fluctuating force vanishes when the average is taken over the total density matrix of the bath degrees of freedom and the coupling. That is

$$\langle f(t)\rangle_{\rho_{(B+SB)}} = \frac{Tr_B[f(t)\exp\left(-\beta(\mathcal{H}_B + \mathcal{H}_{SB})\right)]}{Tr_B[\exp\left(-\beta(\mathcal{H}_B + \mathcal{H}_{SB})\right)]} = 0, \tag{21}$$

where $\rho_{(B+SB)}$ is the shifted canonical equilibrium distribution of the heat bath which is given by

$$\rho_B = Z_B^{-1}\exp\left[-\beta\sum_j\left(\frac{p_j(0)^2}{2m_j} + \frac{m_j\omega_j^2}{2}\left(x_j(0) - \frac{C_j}{m_j\omega_j^2}q(0)\right)^2\right)\right], \tag{22}$$

where Z_B is the partition function of the bath oscillators. Also

$$\langle f(t)f(t')\rangle_{\rho_{(B+SB)}} = Mk_BT\gamma(t-t'). \tag{23}$$

This is the fluctuation-dissipation relation. For weak coupling, we seperate the transient term (or the initial slip) $M\gamma(t)q(0)$ which is of second order in the coupling constant C_j and write the random force as [18].

$$f(t) = \zeta(t) - m\gamma(t)q(0). \tag{24}$$

Therefore the generalized Langevin equation takes the form

$$M\ddot{q}(t) + M\int_0^t dt'\gamma(t-t')\dot{q}(t') + V'(q) = \zeta(t) - M\gamma(t)q(0), \tag{25}$$

where

$$\zeta(t) = \sum_j C_j\left(x_j(0)\cos(\omega_j t) + \frac{p_j(0)}{m_j\omega_j}\sin(\omega_j t)\right). \tag{26}$$

The integration in the forward direction of time in Eq. (25) ensures that it breaks the time reversal invariance explicitly thereby introducing irreversibility in the problem. When taken an average of the Eq. (26) with respect to the canonical classical equilibrium density operator of the unperturbed bath

$$\rho_B(0) = Z_B^{-1}\exp\left[-\beta\sum_j\left(\frac{p_j(0)^2}{2m_j} + \frac{m_j\omega_j^2}{2}x_j(0)^2\right)\right], \tag{27}$$

we obtain

$$\langle\zeta(t)\rangle_B = \frac{Tr_B[\zeta(t)\exp(-\beta\mathcal{H}_B)]}{Tr_B[\exp(-\beta\mathcal{H}_B)]} = 0, \tag{28}$$

and the unequal time correlation

$$\langle\zeta(t)\zeta(t')\rangle_{\rho_B(0)} = Mk_BT\gamma(t-t'). \tag{29}$$

From Eq. (20), it is clear that the force operator depends explicitly on the initial conditions of the bath positions and momenta and also on an inhomogeneous slip term $M\gamma(t)q(0)$. Usually, this term is neglected under the Markovian/Ohmic limit, when the friction assumes the ohmic form $\gamma(t) \to 2\gamma\delta(t)$ [18].

Now, we calculate the correlation function of the random force. We may use either $f(t)$ with respect to $\rho_{(B+SB)}$ or equivalently $\zeta(t)$ with respect to ρ_B. Eqs. (20) and (24) gives

$$\langle\zeta(t)\zeta(0)\rangle_B = \sum_{j,l} C_j C_l\left\langle\left(x_j(0)\cos(\omega_j t) + \frac{p_j(0)}{m_j\omega_i}\sin(\omega_j t)\right)x_l(0)\right\rangle_B. \tag{30}$$

At thermal equilibrium, the second moments of the position and momentum of the bath are calculated and yields

$$\langle x_j(0)x_l(0)\rangle_B = \delta_{jl}\frac{\hbar}{2m_j\omega_j}\coth\left(\frac{\hbar\beta\omega_j}{2}\right), \tag{31}$$

$$\left\langle p_j(0)x_l(0)\right\rangle_B = -\frac{i\hbar}{2}\delta_{jl}. \tag{32}$$

Incorporating Eqs. (31) and (32), the noise correlation in Eq. (30) can be expressed as

$$\langle\zeta(t)\zeta(0)\rangle_B = \sum_{j=1}^N \frac{\hbar C_j^2}{2m_j\omega_j}\left[\coth\left(\frac{\hbar\beta\omega_j}{2}\right)\cos(\omega_j t) - i\sin(\omega_j t)\right]. \tag{33}$$

It is to be noted that the noise correlation contains an imaginary part. This is due to the fact that $\zeta(t)$ and $\zeta(0)$ are now operators in quantum mechanics and, in general, do not commute with each other. We have obtained all the relations corresponding to the classical counterparts, here. Now, we need the reduced description of the quantum system alone and to get such an expression we have to eliminate the bath degrees of freedom from the total picture. The formal way of doing that is to introduce the "spectral density" of the bath oscillators which contains all the information about the heat bath. The spectral density is given by

$$J(\omega) = \pi \sum_{j=1}^{N} \frac{C_j^2}{2m_j\omega_j} \delta(\omega - \omega_j), \tag{34}$$

so that the damping kernel takes a form

$$\gamma(t) = \frac{1}{M} \sum_{j=1}^{N} \frac{C_j^2}{m_j\omega_j^2} \cos(\omega_j t) = \frac{2}{M} \int_0^\infty \frac{d\omega}{\pi} \frac{J(\omega)}{\omega} \cos(\omega t). \tag{35}$$

The most widely used form of the spectral density is of the following form

$$J(\omega) = M\gamma\omega, \tag{36}$$

which is called the *ohmic* and it apparently produces a memoryless friction

$$\gamma(t) = 2\gamma\delta(t). \tag{37}$$

But this is an ideal situation. In real physical situations, the spectral density falls off in the $\omega \to \infty$ limit. The above form (cf. Eq. (36)) of the spectral density, in the large ω limit, gives divergences in certain physical quantities like the momentum dispersion. It is then customary to introduce a cut-off to the spectrum with which the specturm vanishes above that cutoff. Such a spectrum is known as a Drude regularized spectral density [1] and is given by

$$J(\omega) = M\gamma\omega \frac{\omega_D^2}{\omega^2 + \omega_D^2}, \tag{38}$$

where ω_D is a cutoff to the spectrum of bath oscillators above which the spectral density vanishes. From Eq. (35), for positive arguments $t > 0$, the damping or memory kernel takes a form

$$\gamma(t) = \gamma\omega_D \exp(-\omega_D t). \tag{39}$$

We need $\tilde{\gamma}(\omega)$, the Fourier transform of Eq. (39), for our forthcoming calculations and is written as

$$\tilde{\gamma}(\omega) = \frac{\gamma\omega_D}{\omega_D - i\omega}, \quad \text{and } \Re[\tilde{\gamma}(\omega)] = \frac{\gamma\omega_D^2}{\omega_D^2 + \omega^2}. \tag{40}$$

The "tilde" sign is used to denote the Fourier transform of a function throughout the chapter. One can still use the terminology "ohmic damping" even if the Eq. (38)

does not hold above a critical frequency, provided all the typical frequencies appearing in the dynamics should be much lower than this critical frequency. In the strict 'Ohmic' limit the generalized Langevin equation becomes memoryless and corresponds to the classical Langevin equation.

3. The model and the calculation of instantaneous power

We discussed the theoretical descriptions to understand how a dissipative system behaves when connected to a heat bath at equilibrium. The quantum system was considered to be not in equilibrium prior to the coupling. Once the coupling is established the quantum system continuously transfers its energy to the equilibrium bath and eventually, the system reaches thermal equilibrium (asymptotically) with the heat bath. Thermal equilibrium is said to have reached when a quantum system explores its phase-space fully. From here onwards we discuss the various energy exchanges that happen in a many-body system at equilibrium. The thermal properties of the quantum system can be calculated by assuming that the entire system-plus-bath arrangement is embedded in an infinitely large environment which provides the working temperature. Therefore, from here onwards we denote the quantum system as a subsystem of the bigger bath. Of course, the heat bath with N oscillator modes can also be considered as a subsystem of the bigger bath. Put it differently, the quantum system of our interest and the heat bath with which it is connected became the constituents of a large environment (**Figure 2**). Now we discuss how the energy exchange processes within this 'envelope'. We show here how the random force balances the energy lost by the quantum subsystem to the heat bath. It is enough to calculate the work done by the random force to that it compensates for the energy lost from the subsystem. Moreover, this work done by the random force is necessary to maintain equilibrium.

To proceed further, we need a working model for the subsystem. We choose the charged oscillator in a magnetic field as our quantum subsystem. Hence our system of

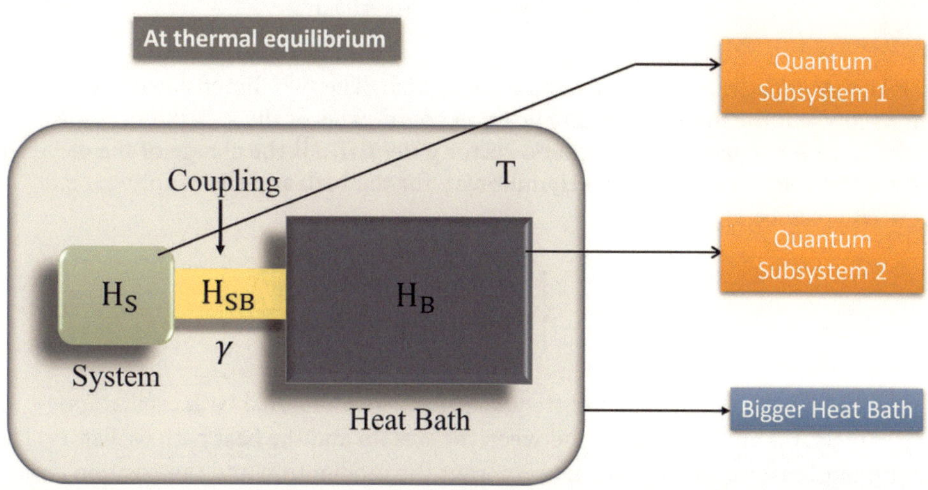

Figure 2.
Pictorial representation of various constituents of a bigger bath. The quantum system as well as the heat bath are now two different parts of the larger environment. The terminology "subsystem" makes better sense in this context.

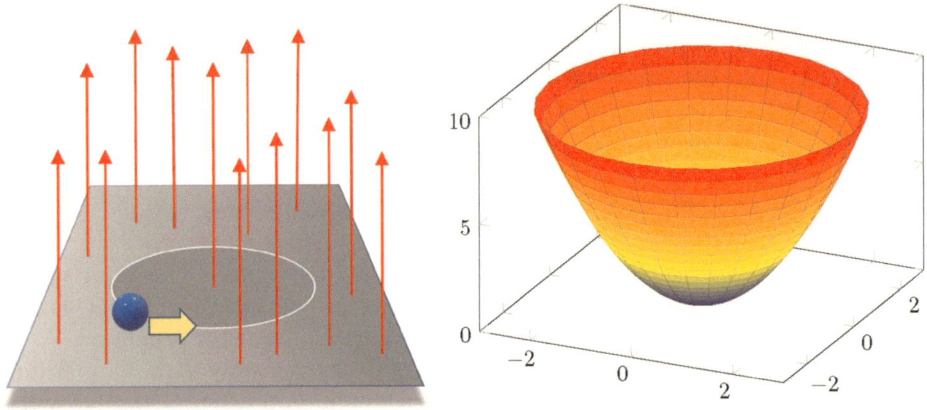

Figure 3.
(a) Electron motion under a perpendicular uniform magnetic field. The trajectories are helical in nature, but the projection of an individual electron's trajectory onto a two-dimensional plane shows a circular motion around the magnetic line of force. (b) The mexican-hat potential. We take this potential to confine the electron under perpendicular magnetic field and the whole arrangement can be simply called a "charged oscillator in a magnetic field". This real physical model is very useful and studied extensively in the condensed matter realm in various contexts. Studies on quantum dots and wires rely on this model heavily.

study is the dissipative charged oscillator in a magnetic field not the uncoupled charged oscillator in a magnetic field. A damped harmonic oscillator was used as a quantum subsystem in the literature [19, 20]. Magnetic field effects in the domain of dissipative quantum physics are of great interest in various phenomena including the quantum Hall effect [21] and superconductivity [22]. Studies on the dissipative charged oscillator in a magnetic field using the system-plus-reservoir approach were originally carried out by Li et al. [23, 24]. This model, later, was used by many others [25–30] in different contexts (**Figure 3**).

$\mathcal{H}_S$ represents the Hamiltonian of the charged magneto-oscillator. It is given by

$$\mathcal{H}_S = \frac{1}{2M}\left(\mathbf{p} - \frac{e\mathbf{A}}{c}\right)^2 + \frac{1}{2}M\omega_0^2\mathbf{r}^2, \tag{41}$$

where M is the mass of the quantum subsystem. The two dimensional vectors $\mathbf{p}$ and $\mathbf{r}$ represent the momentum and position coordinates of the subsystem respectively. Here $\mathbf{A}$ is denotes the magnetic vector potential, e is the charge of the electron, and c is the velocity of light. The Hamiltonian for the bath and the coupling, i.e., $\mathcal{H}_B + \mathcal{H}_{SB}$, can be expressed as

$$\mathcal{H}_B + \mathcal{H}_{SB} = \sum_{j=1}^{N}\left\{\frac{\mathbf{p}_j^2}{2m_j} + \frac{1}{2}m_j\omega_j^2\left(\mathbf{x}_j - \frac{C_j\mathbf{r}}{m_j\omega_j^2}\right)^2\right\}. \tag{42}$$

m_js and ω_js are the masses and frequencies of the individual bath oscillators respectively. C_js are the coupling between the system and the heat bath oscilators. The two dimensional vectors $\mathbf{p}_j$s and $\mathbf{x}_j$s represent the momentum and the position coordinates of the bath oscillators respectively. The position and momentum vectors of the subsystem and the heat bath are operators and they satisfy the following commutation relations

$$\left[\mathbf{r}_i, \mathbf{p}_k\right] = i\hbar\delta_{ik}, \quad \left[\mathbf{x}_{ji}, \mathbf{p}_{lk}\right] = i\hbar\delta_{jl}\delta_{ik}. \tag{43}$$

Following the steps given in the previous section, we use the Heisenberg equations of motion from the total Hamiltonian, we obtain the equations of motion for the subsystem and the bath coordinates. Eliminating the bath variables yields the generalized equation for the system coordinate (which is an operator) and that is given by [30]:

$$M\ddot{\mathbf{r}} + \int_0^t dt'\dot{\mathbf{r}}(t')\gamma(t-t') - \frac{e}{c}(\dot{\mathbf{r}} \times \mathbf{B}) + M\omega_0^2\mathbf{r} = -M\gamma(t)\mathbf{r}(0) + \mathbf{F}(t). \tag{44}$$

This is an initial value equation and has an exact solution as well [30, 31]. Like we saw in the previous section, the spurious initial slip term is here as well. We may define an auxiliary random force $\mathcal{K}(t) = -M\gamma(t)\mathbf{r}(0) + \mathbf{F}(t)$ just for esthetics. The Langevin equation is shown to be independent of any particular choice of gauge since it does not explicitly contain the magnetic vector potential **A**. The appearance of external magnetic field is via the quantum version of the Lorentz force term in the equation. The *memory friction function* $\gamma(t)$ is defined already in the previous section. The stochastic Brownian noise $\mathbf{F}(t)$, depends explicitly on initial coordinates and the momenta of the bath oscillators, is given by

$$\mathbf{F}(t) = \sum_j \left\{ C_j\mathbf{x}_j(0)\cos\left(\omega_j t\right) + \frac{C_j\mathbf{p}_j(0)}{m_j\omega_j}\sin\left(\omega_j t\right) \right\}. \tag{45}$$

It is necessary to note that the damping $\gamma(t)$ and the operator valued Gaussian random force remain unchanged by the external magnetic field. The random force (45) satisfies the following relations [23, 24, 26, 28–30].

$$\begin{aligned} \langle\{F_\alpha(t), F_\kappa(t')\}\rangle &= \delta_{\alpha\kappa}\frac{2}{\pi}\int_0^\infty d\omega \Re[\tilde{\gamma}(\omega + i0^+)]\hbar\omega \\ &\times \coth\left(\frac{\hbar\omega}{2k_BT}\right)\cos\left[\omega(t-t')\right], \end{aligned} \tag{46}$$

$$\begin{aligned} \langle[F_\alpha(t), F_\kappa(t')]\rangle &= \delta_{\alpha\kappa}\frac{2}{i\pi}\int_0^\infty d\omega \Re[\tilde{\gamma}(\omega + i0^+)]\hbar\omega \\ &\times \sin\left[\omega(t-t')\right], \end{aligned} \tag{47}$$

where $\alpha, \kappa = x, y, z$ and $\tilde{\gamma}(z) = \int_0^\infty dt \exp(izt)\gamma(t)$, $(z > 0)$. For the Drude regularized Ohmic spectral density, using Eq. (38), the Eqs. (46) and (47) can be written as

$$\begin{aligned} \langle\{F_\alpha(t), F_\kappa(t')\}\rangle &= \delta_{\alpha\kappa}\frac{2\gamma\omega_D^2}{\pi}\int_0^\infty d\omega \frac{\hbar\omega}{\omega_D^2 + \omega^2} \\ &\times \coth\left(\frac{\hbar\omega}{2k_BT}\right)\cos\left[\omega(t-t')\right], \end{aligned} \tag{48}$$

$$\begin{aligned} \langle[F_\alpha(t), F_\kappa(t')]\rangle &= \delta_{\alpha\kappa}\frac{2\gamma\omega_D^2}{i\pi}\int_0^\infty d\omega \frac{\hbar\omega}{\omega_D^2 + \omega^2} \\ &\times \sin\left[\omega(t-t')\right]. \end{aligned} \tag{49}$$

The anti-symmetric correlation can also be written as

$$\langle\{F_\alpha(t), F_\kappa(t')\}\rangle = 2\delta_{\alpha\kappa}\int_0^\infty d\omega \mathcal{G}(\omega)\cos\left[\omega(t-t')\right], \tag{50}$$

where $\mathcal{G}(\omega)$ is called the power spectrum, and is given by

$$\begin{aligned}\mathcal{G}(\omega) &= \frac{\gamma\omega_D^2}{\pi}\frac{\hbar\omega}{\omega_D^2+\omega^2}\coth\left(\frac{\hbar\omega}{2k_BT}\right)\\ &= \frac{2\gamma\omega_D^2}{\pi\left(\omega_D^2+\omega^2\right)}\left(\frac{\hbar\omega}{2}+\frac{\hbar\omega}{e^{\hbar\omega/k_BT}-1}\right).\end{aligned} \tag{51}$$

The fluctuation dissipation theorem ensures that the symmetric combination and the commutator structure of the random force are (i) proportional to the friction constant γ, and (ii) are independent of the external potential $V(\mathbf{r})$. From Eq. (52), we observe that the power spectrum of the random force is the Planck spectrum, with an extra term due to the zero point fluctuations. Needless to say, in the limit of $\hbar \to 0$ the spectrum becomes the flat spectrum of the white noise.

We now calculate the expectation value of the instantaneous power supplied by the random force. Since we work with operators, we use the symmetric form of the power. The power can be written as

$$\mathcal{P}_F = \frac{1}{2}\langle\mathbf{v}(t)\mathbf{F}(t)+\mathbf{F}(t)\mathbf{v}(t)\rangle, \tag{52}$$

with $\mathbf{v}(t) = \dot{\mathbf{r}}(t)$. We define the variables $Z = x+iy,\quad F = F_x + iF_y$, and $\overline{\gamma}(t) = \frac{\gamma(t)}{m} + i\omega_c$, to re-write the Langevin equation given in Eq. (44) as [30].

$$\ddot{Z} + \int_0^t dt'\overline{\gamma}(t-t')\dot{Z}(t') + \omega_0^2 Z = -\gamma(t)Z(0) + \frac{F(t)}{M}, \tag{53}$$

where $\omega_c = eB/mc$ is the magnetic cyclotron frequency. The new position coordinate is Z and its derivative with respect to time represents the velocity operator. Using Laplace transform and with the aid of fundamental solutions, we find the solution of the Langevin equation given in Eq. (53) as [30, 31].

$$Z(t) = M\dot{\chi}(t)Z(0) + M\chi(t)\dot{Z}(0) + \int_0^t d\tau\chi(t-\tau)F(\tau), \tag{54}$$

where $\chi(t)$ is called the response function of the system and is defined as [30, 31].

$$\chi(t-\tau) = \frac{1}{2\pi i}\int_{-i\infty}^{+i\infty} ds\hat{\chi}(s)\exp\left(s(t-\tau)\right). \tag{55}$$

The Laplace transform of the response function is written as

$$\hat{\chi}(s) = \frac{1}{M}\frac{1}{s^2+\omega_0^2+s\overline{\gamma}(s)}, \tag{56}$$

with the Fourier transform

$$\chi(t-\tau) = \frac{1}{2\pi}\int_{-\infty}^{+\infty} d\omega\tilde{\chi}(\omega)\exp\left(-i\omega(t-\tau)\right), \tag{57}$$

where the dynamical susceptibility is given by

$$\tilde{\chi}(\omega) = \frac{1}{M}\frac{1}{-\omega^2+\omega_0^2-i\omega\overline{\gamma}(\omega)}. \tag{58}$$

We have, here, $\overline{\gamma}(\omega) = i\omega_c + \tilde{\gamma}(\omega)$. The memory kernel $\gamma(t)$ typically falls to zero in the bath relaxation time. As a result, the initial slip term in Eq. (44) (and subsequently in Eq. (53)) vanishes for very long times (can be experimental time scales) compared to the system's characteristic decay time. Therefore, for longer times compared to the system's relaxation time, the quantum Langevin equation (cf. Eq. (53)) becomes a stationary one with the lower limit of integration $-\infty$, i.e., [30]

$$\ddot{Z} + \int_{-\infty}^{t} dt'\overline{\gamma}(t-t')\dot{Z}(t') + \omega_0^2 Z = \frac{F(t)}{M}, \tag{59}$$

with the solution, which is a stationary process, is given by

$$Z(t) = \int_{-\infty}^{t} d\tau\chi(t-\tau)F(\tau). \tag{60}$$

We note from the expressions (Eqs. (53) and (57)) that, so long as ω_0 remains non-zero, the response function $\chi(t-\tau)$ vanishes exponentially for longer times. This, in fact, is in close proximity with the Tauberian theorem which states the asymptotic behavior of a function depends on the low frequency behavior of its Fourier transform. Hence for very longer times, the dependence of $Z(t)$ on the initial values (cf. Eq. (54)) completely disappears and that yield [30, 31].

$$Z(t) = \int_{0}^{t} d\tau\chi(t-\tau)F(\tau). \tag{61}$$

Upon comparing Eq. (61) with Eq. (60), we conclude that $Z(t)$ in Eq. (61) is the solution of the stationary Langevin equation in Eq. (59). With these remarks in mind, we write the Langevin equation in Eq. (44) as

$$M\ddot{\mathbf{r}} + M\int_{-\infty}^{t} dt'\dot{\mathbf{r}}(t')\gamma(t-t') - \frac{e}{c}(\dot{\mathbf{r}}\times\mathbf{B}) + M\omega_0^2\mathbf{r} = \mathbf{F}(t). \tag{62}$$

Using Eqs. (52) and (62), we write the power as

$$\mathcal{P}_F = \frac{d}{dt}\left\langle\frac{1}{2M}\left(\mathbf{P}-\frac{e\mathbf{A}}{c}\right)^2 + \frac{1}{2}M\omega_0^2\mathbf{r}^2\right\rangle + \int_{-\infty}^{t} dt'\gamma(t-t')\frac{1}{2}\langle\dot{\mathbf{r}}(t)\dot{\mathbf{r}}(t') + \dot{\mathbf{r}}(t')\dot{\mathbf{r}}(t)\rangle. \tag{63}$$

Since the Langevin equation describes a stationary process the first term is zero because for a stationary process the expectation values of time dependent quantities must be time translational invariant or, in other words, they are constant. This becomes clear when we evaluate the expectation value of an arbitrary time dependent operator $W(t)$. We write [19].

$$\langle W(t)\rangle = \frac{\mathrm{Tr}\{e^{-\beta H}e^{iHt/\hbar}W(0)e^{-iHt/\hbar}\}}{\mathrm{Tr}e^{-\beta H}} = \frac{\mathrm{Tr}\{W(0)\}}{\mathrm{Tr}e^{-\beta H}}, \tag{64}$$

where $\beta = 1/k_B T$. The cyclic property of the Trace operation is used to get the desired result. Thus it becomes clear that $\langle W(t)\rangle$ is time independent for a canonical ensemble and the derivative of it must be zero. Our task is just to find the second term in Eq. (64). We use Eq. (61) to compute the velocity autocorrelation function. It is to be noted that the real part of the quantity $\frac{1}{2}\left\langle\left\{\dot{Z}(t), \dot{Z}^{\dagger}(t')\right\}\right\rangle$ is equivalent to $\frac{1}{2}\langle\dot{\mathbf{r}}(t)\dot{\mathbf{r}}(t') + \dot{\mathbf{r}}(t')\dot{\mathbf{r}}(t)\rangle$.

We write [30].

$$\frac{1}{2}\langle\{Z(t), Z^{\dagger}(t')\}\rangle = \frac{1}{2m^2}\int_0^t d\tau\int_0^{t'} d\tau'\chi(t-\tau)\chi^*(t'-\tau')\langle\{F(\tau), F^{\dagger}(\tau')\}\rangle, \tag{65}$$

where, using (47), we write [30].

$$\begin{aligned}\langle\{F(\tau), F^{\dagger}(\tau')\}\rangle &= \langle\{F_x(\tau), F_x(\tau')\}\rangle + \langle\{F_y(\tau), F_y(\tau')\}\rangle,\\ &= \frac{4}{\pi}\int_0^{\infty} d\omega\Re[\tilde{\gamma}(\omega+i0^+)]\hbar\omega \times \coth\left(\frac{\hbar\omega}{2k_BT}\right)\cos\left[\omega(\tau-\tau')\right].\end{aligned}$$

After some algebra, we obtain [30].

$$\begin{aligned}\frac{1}{2}\langle\{Z(t), Z^{\dagger}(t')\}\rangle = \frac{\hbar}{M\pi}\int_{-\infty}^{+\infty} d\omega\, \omega\Re\left[\frac{\tilde{\gamma}(\omega)}{m}\right]\tilde{\chi}(\omega)\tilde{\chi}^*(\omega)\\ \times\coth\left(\frac{\hbar\omega}{2k_BT}\right)e^{-i\omega(t-t')},\end{aligned} \tag{67}$$

where

$$\tilde{\chi}(\omega) = -\frac{1}{M}\frac{(\omega+i\omega_D)}{(\omega+i\lambda_1)(\omega+i\lambda_2)(\omega+i\lambda_3)}, \tag{68}$$

Here $\lambda_j s$ are the roots of the cubic equation

$$\omega^3 + i\omega^2(\omega_D + i\omega_c) - \omega\left(\omega_0^2 + \gamma\omega_D + i\omega_c\omega_D\right) - i\omega_0^2\omega_D = 0. \tag{69}$$

Similarly, the complex conjugate $\tilde{\chi}^*(\omega)$ of Eq. (68) is given by

$$\tilde{\chi}^*(\omega) = -\frac{1}{M}\frac{(\omega-i\omega_D)}{(\omega-i\lambda_{1'})(\omega-i\lambda_{2'})(\omega-i\lambda_{3'})}, \tag{70}$$

with the λ_j's are the complex conjugates of λ_js. Since Eqs. (68) and (70) satisfy the following relation

$$M\omega\Re[\tilde{\gamma}(\omega)]\tilde{\chi}(\omega)\tilde{\chi}^*(\omega) = \frac{1}{2i}[\tilde{\chi}(\omega) - \tilde{\chi}^*(\omega)] = \tilde{\chi}''(\omega), \tag{71}$$

we obtain [30].

$$\frac{1}{2}\langle\{Z(t), Z^\dagger(t')\}\rangle = \frac{\hbar}{M\pi}\int_{-\infty}^{+\infty} d\omega\, \tilde{\chi}''(\omega)\coth\left(\frac{\hbar\omega}{2k_BT}\right)e^{-i\omega(t-t')}, \tag{72}$$

where $\tilde{\chi}''(\omega)$ is given by

$$\tilde{\chi}''(\omega) = \frac{1}{M}\frac{\omega\tilde{\gamma}(\omega)}{\left(\omega_0^2 - \omega^2 + \omega\omega_c\right)^2 + \omega^2\tilde{\gamma}^2(\omega)}. \tag{73}$$

Taking the derivatives of Eq. (72) with respect to t and t' yields the velocity autocorrelation

$$\frac{1}{2}\left\langle\left\{\dot{Z}(t), \dot{Z}^\dagger(t')\right\}\right\rangle = \frac{\hbar}{M\pi}\int_{-\infty}^{\infty} d\omega\, \omega^2\, \tilde{\chi}''(\omega)\coth\left(\frac{\hbar\omega}{2k_BT}\right)e^{-i\omega(t-t')}. \tag{74}$$

Substituting the real part of the velocity autocorrelation into the equation for the power yields

$$\begin{aligned}\mathcal{P}_F = &\int_{-\infty}^{t} dt'\gamma(t-t')\frac{\hbar}{M\pi}\int_{-\infty}^{\infty} d\omega\, \omega^2\, \tilde{\chi}''(\omega)\\ &\times\coth\left(\frac{\hbar\omega}{2k_BT}\right)\cos\left[\omega(t-t')\right].\end{aligned} \tag{75}$$

Since the damping $\gamma(t)$ is 0 for negative t and $\tilde{\gamma}(\omega) = \int_0^\infty dt\gamma(t)e^{i\omega t},\ \omega > 0$, the upper limit of the integral in Eq. (44) can be replaced with $+\infty$. Therefore

$$\begin{aligned}\mathcal{P}_F = &\frac{\hbar}{M\pi}\int_{-\infty}^{+\infty} d\omega\int_{-\infty}^{\infty} dt'\ \gamma(t-t')\cos\left[\omega(t-t')\right]\\ &\times\omega^2\,\tilde{\chi}''(\omega)\coth\left(\frac{\hbar\omega}{2k_BT}\right).\end{aligned} \tag{76}$$

It is possible to write

$$\int_{-\infty}^{+\infty} dt'\gamma(t-t')\cos\left[\omega(t-t')\right] = \Re\left\{\int_{-\infty}^{+\infty} dt'\gamma(t-t')e^{i\omega(t-t')}\right\}. \tag{77}$$

Since the bracketed term is just the Fourier transform $\tilde{\gamma}(\omega)$ of $\gamma(t-t')$, we write Eq. (77) as

$$\mathcal{P}_F = \frac{\hbar}{M\pi}\int_{-\infty}^{\infty} d\omega\, \omega^2\, \tilde{\chi}''(\omega)\Re[\tilde{\gamma}(\omega)]\coth\left(\frac{\hbar\omega}{2k_BT}\right). \tag{78}$$

It is immediately observed that only the even part of the above integral contributes. We know that $\tilde{\gamma}(-\omega)^* = \tilde{\gamma}(\omega)$ and $\tilde{\chi}(-\omega) = \tilde{\chi}(\omega)$, so that $\tilde{\chi}''(\omega)$ is an odd function of ω and $\Re[\tilde{\gamma}(\omega)]$ is an even function of the frequency ω. Upon using the real part of $\tilde{\gamma}(\omega)$, we write

$$\mathcal{P}_F = \frac{\hbar}{M\pi}\int_{-\infty}^{\infty} d\omega\, \omega^2\, \tilde{\chi}''(\omega)\frac{\gamma\omega_D^2}{\omega_D^2+\omega^2}\coth\left(\frac{\hbar\omega}{2k_BT}\right). \tag{79}$$

This is the instantaneous power supplied by the random force for our particular model and the above expression is positive quantity always.

To check the result, we evaluate the power supplied by the random force at high temperatures (or in the classical case). After some tedious mathematical manipulations we obtain [32].

$$\mathcal{P}_F = \frac{4k_BT\gamma}{M}\frac{\left(2+\omega_0^2/\omega_D^2+\gamma/\omega_D\right)}{\left[\left(2+\omega_0^2/\omega_D^2+\gamma/\omega_D\right)^2+\left(2\omega_c/\omega_D\right)^2\right]}, \tag{80}$$

which in the large cutoff (strict ohmic) limit ($\omega_D \to \infty$) yields the form

$$\mathcal{P}_F = \frac{4k_BT\gamma}{M}. \tag{81}$$

This clearly indicates to us that the rate of work done by the random force on the quantum subsystem is indeed proportional to the damping/dissipation strength γ. This in turn tells us that, at equilibrium, the energy lost by the quantum subsystem due to dissipation is compensated by an amount of energy received from the random force. As a consequence of the fluctuation-dissipation theorem, the Eq. (81) does not contain any term from the external potential (and magnetic field), in the strict ohmic limit.

4. Discussion of the result

The full quantum many-body system has an infinite number of degrees of freedom each with its corresponding zero-point oscillations. The full system must be in the ground state at the absolute zero of temperature ($T = 0$). This further implies that no work is done on or by the system. However, for any finite coupling (irrespective of the strength of the coupling), $\mathcal{H}_S$ does not commute with $\mathcal{H}$. That means the ground state of $\mathcal{H}$ is not the ground state of $\mathcal{H}_S$. Therefore, even at absolute zero temperature, the energy of the charged oscillator in a magnetic field must fluctuate. Mathematically, this statement can be expressed as

$$\Delta\mathcal{H}_S^2 = \langle\mathcal{H}_S^2\rangle - \langle\mathcal{H}_S\rangle^2 \neq 0, \tag{82}$$

the mean square fluctuations of $\mathcal{H}_S$ is not equal to zero. No matter how weak the coupling between the subsystem and the bath, the fluctuation of the subsystem Hamiltonian does not vanish at $T = 0$. This fluctuation is obviously driven by the random force. Therefore, the work done per unit time by this random force is, indeed, balanced by the dissipative loss of the subsystem. Hence there is no net work done on the subsystem. To evaluate the RHS of Eq. (82) we resort to Eq. (41). We write

$$\langle \mathcal{H}_S \rangle = \frac{1}{2M}\left\langle \left(\mathbf{p} - \frac{e\mathbf{A}}{c}\right)^2 \right\rangle + \frac{1}{2}M\omega_0^2\langle \mathbf{r}^2 \rangle, \tag{83}$$

In the new variable $Z = x + iy$, the subsystem Hamiltonian can be written as [30].

$$\mathcal{H}_S = \frac{1}{2}M\dot{Z}\dot{Z}^\dagger - \frac{1}{2}\hbar\omega_c + \frac{1}{2}M\omega_0^2 ZZ^\dagger. \tag{84}$$

Therefore

$$\langle \mathcal{H}_S \rangle = \frac{1}{2}M\left\langle \dot{Z}\dot{Z}^\dagger \right\rangle - \frac{1}{2}\hbar\omega_c + \frac{1}{2}M\omega_0^2\langle ZZ^\dagger \rangle. \tag{85}$$

We need to compute the unequal time correlation functions $\langle Z(t)Z^\dagger(t') \rangle$ and $\left\langle \dot{Z}(t)\dot{Z}^\dagger(t') \right\rangle$. To get this, we only evaluate the symmetric combination of $Z(t)$ and $Z^\dagger(t')$. The anti-symmetric combination vanishes in the analytic continuation $t' = t$ to obtain the equilibrium values [30].

Using Eqs. (72) and (74), at $T = 0$, we compute Eq. (85). We get [30].

$$\begin{aligned} \langle \mathcal{H}_S \rangle_{T=0} = {} & \frac{\hbar\gamma}{\pi}\ln\left(\frac{\omega_D}{\omega_0}\right) + \frac{\hbar\omega_0^2}{\pi}\left\{\frac{1}{\Lambda}\ln\left(\frac{\lambda_1}{\lambda_2}\right) + \frac{1}{\Lambda'}\ln\left(\frac{\lambda_1'}{\lambda_2'}\right)\right\} \\ & - \frac{\hbar}{4\pi}\left\{\frac{\overline{\gamma}^2}{\Lambda}\ln\left(\frac{\lambda_1}{\lambda_2}\right) + \frac{\overline{\gamma}^{*2}}{\Lambda'}\ln\left(\frac{\lambda_1'}{\lambda_2'}\right)\right\}, \end{aligned} \tag{86}$$

where $\Lambda = \sqrt{\overline{\gamma}^2 - 4\omega_0^2}$ and $\Lambda' = \sqrt{\overline{\gamma}^{*2} - 4\omega_0^2}$. Also $\lambda_{1,2} = \overline{\gamma}/2 \pm i\sqrt{\omega_0^2 - \overline{\gamma}^2/4}$ and $\lambda_{1,2}' = \overline{\gamma}^*/2 \mp i\sqrt{\omega_0^2 - \overline{\gamma}^{*2}/4}$. For weak dissipation $\overline{\gamma}, \overline{\gamma}^* \ll 2\omega_0$,

$$\begin{aligned} \langle \mathcal{H}_S \rangle_{T=0} = {} & \frac{\hbar\gamma}{\pi}\ln\left(\frac{\omega_D}{\omega_0}\right) - \frac{\hbar\gamma\omega_c}{2\pi\omega_0}\ln\left\{\frac{(\overline{\gamma}/2 + i\omega_0)(\overline{\gamma}^*/2 - i\omega_0)}{(\overline{\gamma}/2 - i\omega_0)(\overline{\gamma}^*/2 + i\omega_0)}\right\} \\ & + \frac{i\hbar(\gamma^2 + \omega_c^2 - 4\omega_0^2)}{4\pi\omega_0}\ln\left\{\frac{(\gamma^2 + \omega_c^2 - 4\omega_0^2) + 4i\gamma\omega_0}{(\gamma^2 + \omega_c^2 - 4\omega_0^2) - 4i\gamma\omega_0}\right\}. \end{aligned} \tag{87}$$

In the absence of the magnetic field ($\omega_c = 0$) we get

$$\langle \mathcal{H}_S \rangle_{T=0} = \hbar\omega_0 + \frac{\hbar\gamma}{\pi}\ln\left(\frac{\omega_D}{\omega_0}\right), \tag{88}$$

which is nothing but the result for a two dimensional isotropic oscillator. Since $\mathbf{r}$ and $\dot{\mathbf{r}}$ (or Z and $\dot{Z}$) are Gaussian operators with zero mean values, $\langle \mathcal{H}_S^2 \rangle$ (cf, Eq. (41)) can be readily expressed in terms of products of $\langle \mathbf{r}^2 \rangle$ and $\langle \dot{\mathbf{r}}^2 \rangle$ (or using Z^2 and $\dot{Z}^2$ with Eq. (84)). We evaluate the $\langle \mathcal{H}_S^2 \rangle$ in the absence of magnetic field to just show the concerned point. Magnetic field has no significant role in the present evaluation to show that the mean square fluctuation of the subsystem Hamiltonian does not vanish. With the aid of Eq. (88), we write

$$\Delta\mathcal{H}_S^2 = \langle \mathcal{H}_S \rangle \frac{\hbar\gamma}{\pi}\ln\left(\frac{\omega_D}{\omega_0}\right). \tag{89}$$

Even for a very weak damping, the mean of the subsystem energy is above its ground state energy and that the fluctuations in this energy do not vanish. This tells us that one part of a physical system in its ground state can and does exchange energy with another part. The formalism we have chosen to show the energy balance is an exact one. We use the Langevin equation to calculate the expectation values after taking the time derivative of the subsystem Hamiltonian.

$$\frac{d}{dt}\langle\mathcal{H}_S\rangle + \gamma\langle\dot{\mathbf{r}}^2(t)\rangle = \frac{1}{2}\langle\dot{\mathbf{r}}(t)\mathbf{F}(t) + \mathbf{F}(t)\dot{\mathbf{r}}(t)\rangle. \tag{90}$$

The exact equation given above is obtained under the strict ohmic limit of the Langevin Equation in Eq. (62). The expectation values appearing above are equal time expectation values and are independent of time as the total physical system is invariant under the time translation. The first term in the above relation is zero. The rest of the equation then describe that the power (RHS) is actually proportional to the dissipation constant γ. Power does not vanish even at absolute zero temperature and this indicates that different parts of the physical system can and do exchange energy even at absolute zero temperature.

Author details

Jishad Kumar
Department of Chemical and Biological Physics, Weizmann Institute of Science, Rehovot, Israel

*Address all correspondence to: kumar.jishad@gmail.com

References

[1] Weiss U. *Quantum Dissipative Systems*. 3rd ed. Singapore: World Scientific; 2008

[2] Dittrich T, Hänggi P, Ingold G-L, Kramer B, Schön G, Zwerger W. *Quantum Transport and Dissipation*. New York, NY: Wiley-VCH; 1998

[3] Nakajima S. Progress in Theoretical Physics. 1958;**20**:948

[4] Zwanzig R. The Journal of Chemical Physics. 1960;**33**:1338

[5] Zwanzig R. In: Brittin WE, Downs BW, Down J, editors. *Lectures in Theoretical Physics*. Vol. 3. Boulder, New York: Interscience; 1961; Also see, Zwanzig R. Nonequilibrium Statistical Mechanics. New York: Oxford University Press; 2001

[6] Prigogine J, Resibois P. Physica. 1961; **27**:629

[7] Senitzky IR. Physics Review. 1961; **119**:670; **124** 642

[8] Ford GW, Kac M, Mazur P. Journal of Mathematical Physics. 1965;**6**:504

[9] Mori H. Progress in Theoretical Physics. 1965;**33**:423

[10] Caldeira AO, Leggett AJ. Physical Review Letters. 1981;**46**:211

[11] Caldeira AO, Leggett AJ. Annals of Physics (NY). 1983;**149**:374

[12] Ford GW, Lewis JT, O'Connell RF. Physical Review A. 1988;**37**:4419

[13] Ullersama P. Physica. 1966;**32**:27, 56, 74, 90

[14] Magalinski VB. Zhurnal Eksperimental'noi i Teoreticheskoi Fiziki. 1959;**36**:1942 [Sov. Phys. JETP **9** 1381]

[15] Zwanzig R. Journal of Statistical Physics. 1973;**9**:215

[16] Grabert H, Schramm P, Ingold G-L. Physics Reports. 1988;**168**:115

[17] Hemmer PC, Maximon LC, Wergeland H. Physics Review. 1958;**111**: 689

[18] Hänggi P. Lecture Notes in Physics. 1997;**484**:15-22

[19] Li XL, Ford GW, O'Connell RF. Physical Review E. 1993;**48**:1547

[20] Li XL, Ford GW, O'Connell RF. Physical Review E. 1995;**51**:5169

[21] Prange RE, Girvin SM, editors. The Quantum Hall Effect. Berlin: Springer; 1987

[22] Ong NP. In: Ginsberg DM, editor. *Physical Properties of High Temperature Superconductors*. Vol. 2. Singapore: World Scientific; 1990

[23] Li XL, Ford GW, O'Connell RF. Physical Review A. 1990;**41**:5287

[24] Li XL, Ford GW, O'Connell RF. Physical Review A. 1990;**42**:4519

[25] Hong TM, Wheatley JM. Physical Review B. 1991;**42**:6492; **43** 5702

[26] Li XL, O'Connell RF. Physica A. 1996;**224**:639-668

[27] Li XL, Ford GW, O'Connell RF. Physical Review E. 1996;**53**:3359

[28] Dattagupta S, Singh J. Pramana. 1996;**47**:211

[29] Dattagupta S, Singh J. Physical Review Letters. 1997;**79**:961

[30] Kumar J. Physica A. 2014;**393**: 182-206

[31] Ford GW, O'Connell RF. Physical Review D. 2001;**64**:105020

[32] Kumar J. Physica Scripta. 2014;**89**: 095701

Chapter 4

Centrifugal Acceleration in Relativistic Astrophysics

Andria Rogava

Abstract

Particles moving along prescribed, relativistically rotating trajectories may exhibit quite unexpected, interesting kinematic behavior. Their dynamics may lead to a number of physical processes that could have various important consequences in a wide variety of relativistic astrophysical objects. In this chapter the author gives brief review of the theoretical ideas and results related to Machabeli and Rogava *gedankenexperiment* (1994) and their astrophysical implications. In particular three astrophysical cases—acceleration of particles by rotating magnetospheres in AGNs, centrifugal acceleration and gamma flares in Crab nebula, and self-trapping as a beaming mechanism for Fast Radio Bursts—are discussed. Conclusions and future research prospects are briefly outlined.

Keywords: special relativity, fundamental problems and general formalism, relativity and gravitation, centrifugal acceleration, relativistic astrophysics

1. Introduction

Crisis is a Hair
Toward which forces creep
Past which forces retrograde ... .
- Emily Dickinson.

Christiaan Huygens introduced the concept of the centrifugal force in the manuscript entitled "*De vi Centrifuga*" written in 1659 [1] and published posthumously in Leiden, in 1703. Since then, the concept of the centrifugal force became an integral part of the essential vocabulary of physics and a topic of different interpretations in the philosophy of the physical science. It proved to be a useful notion for studies of a wide variety of problems involving rotation and dynamics of particles/bodies in rotating frames of reference. Rotation is omnipresent in the nature; as long as the humans advanced in the space exploration, it became increasingly evident that the physical processes related to the presence of rotation play an important, often decisive, role in the complex and still incompletely understood texture of the physical universe.

From the very beginning, there was a certain kind of dichotomy in the understanding of the centrifugal force. According to Newton and basic principles of the classic mechanics, the magnitude and the direction of the centrifugal force, acting on matter in a rotating frame of reference, are not connected with motion of the matter, but are entirely and explicitly determined by the rotation of the reference frame. On the basis of this postulate, the centrifugal force is defined in terms of the angular velocity of the instantaneously corotating reference frame and a displacement of the moving body (particle) from the instantaneous rotation axis. Consequently, the direction of the force is always unequivocally determined by the direction of the displacement vector and it always points outward. However, it has to be remembered that the original concept of the centrifugal force, outlined by Huygens [2], emphasized another, drastically different, aspect of the force. In particular, Huygens' idea was the following: "*centrifugal force should not be connected to rotation of a reference frame, but to motion along a curved path in space.*" Within the framework of classic, nonrelativistic physics, the difference between these two postulates does not lead to a quantitative difference, because in the Newtonian gravitation theory, based on Euclidean geometry, these two definitions are equivalent. But in Relativity, when the space–time is not Euclidean, they differ from each other [2] and may lead to different results.

In recent decades, the increase of interest in the phenomenon of relativistic rotation was largely due to series of significant results by M. Abramowicz and his collaborators [2–5] related to physics of relativistic rotation in the vicinity of black holes. In particular, it was found that under certain circumstances, the centrifugal force attracts toward the rotation axis even in the case of a nonrotating Schwarzschild black hole. This outstanding fact led to a number of unusual rotational effects considered and discussed in these publications.

About 30 years ago, Machabeli and Rogava considered [6] a "*gedankenexperiment*" (viz. an experiment carried out in thought only): centrifugally driven motion of a bead inside a long, straight pipe rotating with a constant angular velocity $\omega = const$ around an axis normal to its symmetry axis. The motion of the bead was studied in the frame of reference of the rotating pipe. Different classes of exact analytic solutions were found, exhibiting puzzling regimes of motion: it was discovered that centrifugally driven bead could both accelerate and decelerate, and the acceleration could change its sign; actual modes of motion strongly depended on the value of the initial velocity. The purpose of the idealized "*gedankenexperiment*" was to mimic the motion of centrifugally driven charged particles in various astronomical situations. The possible importance of this result in the context of centrifugally driven dynamics of particles in rapidly rotating pulsar magnetospheres and astrophysical jets was later repeatedly pointed out [7] and extensively studied in a number of subsequent publications [8–12]. Namely, particle acceleration by rotating magnetospheres in active galactic nuclei was investigated [13]; the role of radiation reaction forces in the dynamics of centrifugally accelerated particles was explored [14]; centrifugal acceleration was studied in isotropic photon fields [15] and in wormhole metrics [16].

In this chapter of the present book, I will try to give a brief but systematic overview of the Machabeli & Rogava *gedankenexperiment* and its astrophysical implications. The material is divided into two sections: the next one is dedicated to the theoretical ideas and results, while the final one deals with astrophysical applications of the theory. The latter also contains a very short description of unresolved issues and possible directions for further, both physical and astrophysical, research.

2. Theoretical background

In this section, a brief overview of theoretical ideas and results related to the centrifugally driven motion of particles along prescribed trajectories is given. Firstly, main characteristics of the original *gedankenexperiment* [6] are derived and discussed in detail. Secondly, it is shown how the idea developed and what is the range of related physical results disclosed in the process.

2.1 Original gedankenexperiment: Rotating frame study

In Ref. [6], a straight pipe rotating around an axis normal to the pipe, and a small bead moving inside the pipe without the friction were considered; radii of the bead and the pipe being equal to one other. According to Newtonian mechanics the bead would move with ever-increasing acceleration, if at the moment $t = 0$ it was located just above the axis pivot ($r_0 = 0$) and had an initial velocity $v_0 \neq 0$, then the solution of Newton's second law of motion would imply that the radial distance $r(t)$ from the axis would increase in time as:

$$r(t) = (v_0/\omega)\, sinh\,(\omega t), \tag{1}$$

while the bead's radial velocity $v_r(t)$ along the pipe would monotonously increase as:

$$v_r(t) = v_0\, cosh\,(\omega t). \tag{2}$$

However, relativity-related common sense tells us that if the pipe is long enough and its walls are absolutely rigid, then the bead's increasing velocity sooner or later would become relativistic, $v \ll c$ condition sooner or later ceases to hold and the bead's motion has to be scrutinized by means of special relativity. Intuitively it seems evident that the increase of the velocity would somehow be limited since the total velocity of the bead $v_{tot} \equiv \sqrt{v_r^2 + r^2\omega^2}$ cannot exceed the speed of light. Therefore, the problem has to be considered within the framework of a relativistic theory. Hereafter, we shall use *geometrical units*, in which $G = c = 1$. If the bead would reach the *light cylinder* (defined as the radial distance $r^* \equiv 1/\omega$, that is, $v_\phi(t) \equiv \omega r^* = 1$) its radial velocity at r^* should fall to zero value, implying that the radial velocity $v_r(t)$, being increasing at the initial stage of the motion, must become decreasing as the bead approaches the light cylinder.

In Ref. [6] the problem was considered in a reference frame of the rotating pipe (rotating frame, hereafter referred to as RF), where the dynamics of the bead motion appeared to be the simplest: one-dimensional, radial motion along the straight pipe. If the space–time in the laboratory frame (hereafter referred to as LF) is Minkowskian with the metric:

$$ds^2 \equiv -d\tau^2 = -dT^2 + dX^2 + dY^2, \tag{3}$$

then applying the transformation of variables: $T = t, X = rcos(\omega t), Y = rsin(\omega t)$ it can be rewritten in the following form:

$$ds^2 \equiv -d\tau^2 = -(1 - \omega^2 r^2)dt^2 + dr^2, \tag{4}$$

where τ is a proper time of the moving bead, defined in the standard way. The Lagrangian for the derived "1 + 1" space–time metric (4) may be defined as [17]:

$$L \equiv \frac{g_{\alpha\beta}}{2}\dot{x}^\alpha \dot{x}^\beta = \frac{1}{2}\left[-(1 - \omega^2 r^2)\left(\frac{dt}{d\tau}\right)^2 + \left(\frac{dr}{d\tau}\right)^2\right]. \tag{5}$$

Components of the corresponding Lagrange equation:

$$\frac{d}{d\tau}\left(\frac{\partial L}{\partial \dot{x}^{\alpha}}\right) = \frac{\partial L}{\partial x^{\alpha}}, \tag{6}$$

may be written in the following way:

$$-\left(1-\omega^2 r^2\right)\left(\frac{dt}{d\tau}\right) = const \equiv -E, \tag{7}$$

$$\frac{d^2 r}{d\tau^2} = \omega^2 r\left(\frac{dt}{d\tau}\right)^2. \tag{8}$$

A simpler equation for $d^2r/d\tau^2$ may be derived by combining (7) with the algebraic relation between 4-velocities, resulting from their normalization $g_{\alpha\beta}u^{\alpha}u^{\beta} = -1$ condition:

$$-\left(1-\omega^2 r^2\right)(dt/d\tau)^2 + (dr/d\tau)^2 = -1. \tag{9}$$

The result of a simple calculation is:

$$\left(\frac{dr}{d\tau}\right)^2 = -1 + \frac{E^2}{1-\omega^2 r^2}. \tag{10}$$

On the other hand, rewriting (9) as

$$\left(\frac{dt}{d\tau}\right)^2 = \left[1-\omega^2 r^2 - \left(\frac{dr}{dt}\right)^2\right]^{-1}, \tag{11}$$

and expressing from (7) $dt/d\tau$ by means of E, we can write down the following first-order differential equation for the function $r(t)$:

$$drdt = \sqrt{\left(1-\omega^2 r^2\right)\left[1-\frac{\left(1-\omega^2 r^2\right)}{E^2}\right]}. \tag{12}$$

A constant parameter $E = -U_t$, coming into view in these equations, may be treated as an energy of the moving bead in the RF and may be determined through initial conditions. For the above-mentioned case, when at the moment $t = 0$, the bead was at the position $r = r_0$ and had a velocity $v_r = v_0$, we can easily find out that

$$E = \frac{\left(1-\omega^2 {r_0}^2\right)}{\sqrt{1-\omega^2 {r_0}^2 - {v_0}^2}}. \tag{13}$$

Obviously, $0 < E < \infty$. If, for example, $r_0 = 0$ and $v_0 \neq 0$ then $E = \left(1-v_0^2\right)^{-1/2}$ and is more than unity ($E \gg 1$, if $v_0 \approx 1$). While, if $v_0 = 0$ and $r_0 \neq 0$ then we see that $E = \left(1-\omega^2 r_0^2\right)^{1/2}$, certainly, is less than unity. Note also that $E \approx 0$ if $\omega r_0 \approx 1$.

It must be noted that in the nonrelativistic limit the energy of the moving bead reduces to the following expression:

$$E_{nr} = 1 + v^2/2 - \omega^2 r^2/2, \tag{14}$$

where a unity evidently describes rest mass energy (per unit mass) of the bead, while remaining terms have clear "nonrelativistic" physical meaning. In particular, $v^2/2$ is a kinetic energy corresponding to radial motion in the RF, while $-\omega^2 r^2/2$ is a "centrifugal energy" [18], known from classical mechanics.

One can also derive a differential equation for a radial acceleration of the bead d^2r/dt^2. The result is:

$$d^2rdt^2 = \frac{\omega^2 r}{1 - \omega^2 r^2}\left[1 - \omega^2 r^2 - 2\left(\frac{dr}{dt}\right)^2\right], \tag{15}$$

at its left-hand side appears a radial acceleration of the bead as measured in the RF. Note that in nonrelativistic limit, it reduces to the conventional expression for the centrifugal force$f_{cf} = w^2 r$.

In Ref. [6] it was shown that the exact analytic solution of (12) may be expressed by means of elliptical functions. It was derived by introducing two auxiliary variables $\theta \equiv arccos\,(\omega r)$ and $\lambda \equiv \omega t$; and a parameter $m \equiv 1/E^2$, reducing (12) to:

$$\frac{d\theta}{d\lambda} = -\sqrt{1 - msin^2\theta}. \tag{16}$$

The solution of this equation may be written as:

$$\lambda = \int_0^{\varphi_0} \frac{d\theta}{\sqrt{1 - msin^2\theta}} - \int_0^{\varphi} \frac{d\theta}{\sqrt{1 - msin^2\theta}}, \tag{17}$$

with $\varphi_0 = arccos\,(\omega r_0)$. It can be rewritten as:

$$\varphi = am(\lambda^* - \lambda), \tag{18}$$

where *am* is an amplitude of Jacobian elliptic functions [19] and we have introduced an additional parameter:

$$\lambda^* \equiv \int_0^{\varphi_0} \frac{d\theta}{\sqrt{1 - msin^2\theta}}. \tag{19}$$

Recovering physical variables t and r one comes to the following explicit solution:

$$r(t) = \frac{1}{\omega} cn(\lambda^* - \omega t) \tag{20}$$

where cn is Jckobian elliptic cosine [19].

It is the *general* solution to the problem. In Ref. [6] a particular case was considered, specified by above-mentioned initial conditions: $r = 0$ and $v_0 \neq 0$ at $t = 0$. For the case, $m = 1 - v_0^2$ and $\lambda^* = K$, where K is a *complete elliptical integral of the first kind* [19]. It allows rewriting the general solution (20) in a different way:

$$r(t) = \frac{1}{\omega} cn(K - \omega t) = \left(\frac{v_0}{\omega}\right)\frac{sn(\omega t)}{dn(\omega t)}, \tag{21}$$

where *sn* and *dn* are Jacobian elliptical sine and modulus, respectively [19]. The corresponding solution for the radial velocity $v_r(t)$ is:

$$v_r(t) = v_0 \frac{cn(\omega t)}{dn^2(\omega t)}. \tag{22}$$

Both asymptotic limits—nonrelativistic and ultrarelativistic—of the solutions (21) and (22) are easy to specify. When an initial radial velocity of the bead is nonrelativistic ($v_0 \ll 1$), $m \approx 1$ and recalling an asymptotic behavior of Jacobi's elliptic functions [19] ($sn \to tanh, cn \to sech, dn \to sech$) we can see that the solutions reduce to abovementioned (1) and (2) Newtonian expressions. Whereas in the ultrarelativistic limit ($v_0 \approx 1$), $m \approx 0$, hence $sn \to sin, cn \to cos, dn \to 1$ and consequently $r(t) \approx (v_0/\omega) sin(\omega t)$, and $v_r \approx v_0 cos(\omega t)$.

Figure 1 is taken from Ref. [6]. It shows the temporal evolution of the radial velocity of the bead v_r (dashed line) and the quantity $v_\varphi \equiv \omega r$ (solid line), which is equal to the azimuthal velocity of the bead in the laboratory frame. The curves correspond to four different values of v_0. Namely, **Figure 1(a)** is drawn for the case $v_0 = 0.001$. Initially, both velocities (v_r and v_φ) grow almost synchronously, but later radial velocity slows its increasing pace, it reaches its maximum value ($v_{max} \approx 0.5$) and begins to decrease, while v_φ continues increasing and reaches $v_\varphi = 1$ value at the moment $t^* = K/\omega$, when the bead is at the distance $r = r^*$ from the rotation axis and its radial velocity becomes zero. This is a "turning point" since here $v_r(t)$ changes its sign, and the bead begins to move toward the rotation axis with increasing speed. A modulus of the radial velocity in the time interval $t^* < t < 2t^*$ varies exactly in the same way as in the previous interval $0 < t < t^*$. At the moment $t = 2t^*$ the bead is at its starting point $r = 0$ just above the rotation axis and it has the same velocity v_0 but is directed, this time, in the opposite direction. Therefore, the bead passes over the

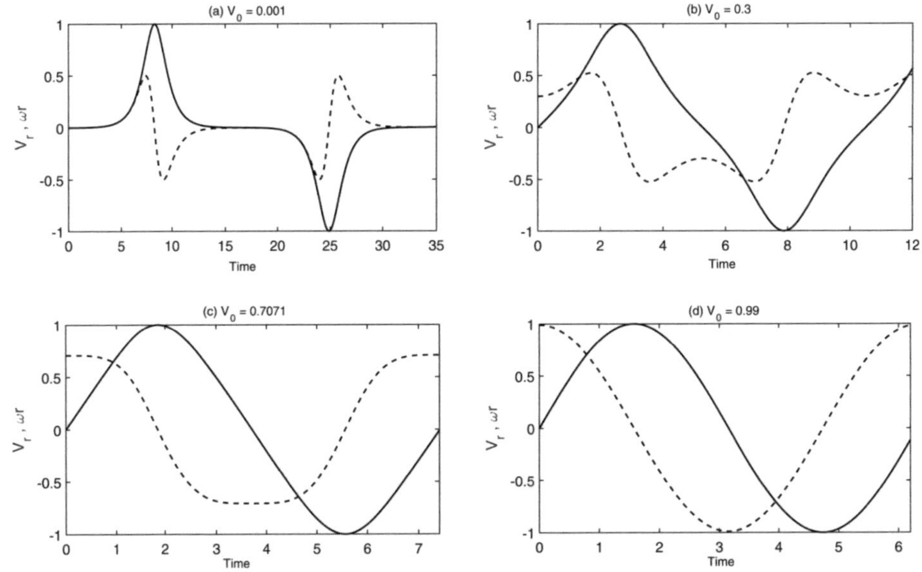

Figure 1.
Dependence of the bead's radial velocity v_r *(dashed line) and azimuthal velocity* $v_\phi \equiv \omega$ r *(solid line) for four different values of initial velocity: (a)* $v_o = 0.001$, *(b)* $v_o = 0.3$, *(c)* $v_0 = \sqrt{2}/2$, *and (d)* $v_o = 0.99$.

rotation axis and, in the interval, $2t^* < t < 4t^*$ repeats in the left half of the pipe the same kind of motion. **Figure 1(b)** is drawn for the case $v_0 = 0.3$. Qualitatively it resembles the previous case. However, a maximum value of v_r is slightly larger and an average radial velocity of the motion is apparently higher.

An interesting "threshold" case $v_0 = \sqrt{2}/2$ is represented in **Figure 1(c)**. From (15) we can see that for this particular case, initial radial acceleration is equal to zero. Therefore, originally, the bead moves almost uniformly, but further on it continues to move with a decreasing radial speed. Other qualitative features of the motion remain the same as in above-specified cases. Finally, **Figure 1(d)** represents the case of a strongly relativistic initial speed $v_0 = 0.99$. It corresponds to the asymptotic case $m \approx 0$ discussed above and consequently the curves for v_r and v_φ would be well represented by usual trigonometric cosine and sine, respectively.

Thus, we see that the character of the bead motion is "oscillatory." The period of the "oscillations" $P \equiv 4t^* = 4K/\omega$. The period tends to infinity when $v_0 \to 0$ (as it should be), while when $v_0 \to 1$, the period naturally tends to: $P \to 2\pi/\omega$.

In Ref. [6] the authors introduced the concept of an "effective potential" $U(r)$, which turns out to be quite useful for shedding more light on the qualitative characteristics of derived solutions. Substituting dr/dt (15) from (12) it is possible to rewrite the equation of motion as:

$$\frac{d^2r}{dt^2} = \omega^2 r\left[(2m-1) - 2m\omega^2 r^2\right] \equiv -\frac{dU(r)}{dr}, \tag{23}$$

and for our particular case ($r_0 = 0, v_0 \neq 0$), taking into account (13), we can rewrite it in the following way:

$$\frac{d^2r}{dt^2} = \omega^2 r\left[\left(1-2v_0^2\right) - 2\left(1-v_0^2\right)\omega^2 r^2\right], \tag{24}$$

while the explicit expression for $U(r)$ is:

$$U(r) = \frac{\omega^2 r^2}{2}\left[(1-2m) + m\omega^2 r^2\right]. \tag{25}$$

In **Figure 2**, which is also taken from [6], the function $U(x)(x \equiv \omega r)$ is shown for four different values of v_0, corresponding to the cases shown in **Figure 1(a–d)**. It provides graphic illustration of the bead's motion in terms of its motion in the specific kind of potential "well." Namely, in (a) and (b) cases, the curves have "secondary" minima, ascertaining that the bead beginning motion from the point $x = 0$ accelerates while "rolling down" to the "secondary" minimum point, then hampers until it reaches the point $x = 1$, and stops and begins to move toward the rotation axis. Case (c) is represented by a "potential well" with an almost flat bottom, where initially, as we have seen earlier, motion happens to be almost uniform. For the fourth case, the form of the "well" naturally implies the motion with the negative radial acceleration during the whole course of the motion.

Recently the *gedankenexperiment* in the RF was reconsidered by Khomeriki and Rogava [20] and the exact solutions for the bead motion were found and analyzed for most general sets of initial conditions. The qualitative nature of the motion is largely the same, but there are interesting quantitatively new classes of solutions. We address the interested reader to this publication for more details.

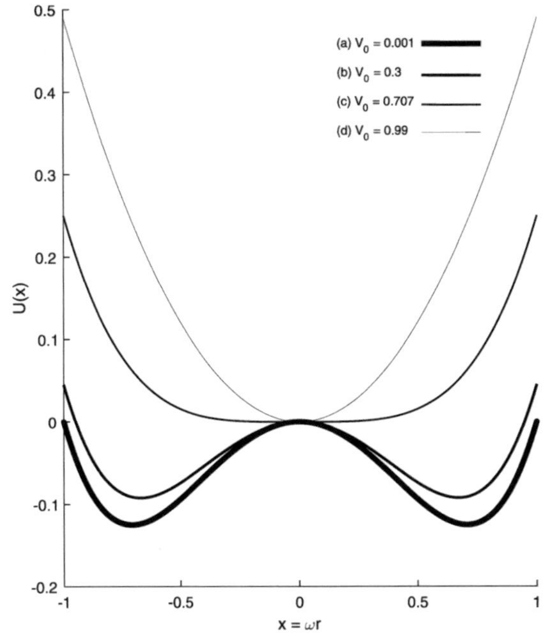

Figure 2.
Effective potential as a function of the quantity $v_\phi \equiv \omega$ r *for four different values of the initial velocity: (a)* v_o = *0.001, (b)* v_o = *0.3, (c)* $v_o = \sqrt{2}/2$, *and (d)* v_o = *0.99.*

It must be noted that in Ref. [6], Gia Machabeli and I have interpreted the surprising behavior of the bead in terms of reversal in direction of a centrifugal force. Criticizing that interpretation, Miller and Abramowicz [21] pointed out the connection of the bead's behavior with relativistic dependence of its mass on velocity. They recommended defining the relativistic centrifugal force in the way that excludes the "reversal" of the force and argued that the deceleration of the bead has to be ascribed to the relativistic mass variation. In Ref. [22], F. de Felice considered relativistic dynamics on a rotating disk and for the case of radially constrained motion, he interpreted the vanishing of the bead's radial velocity at the light cylinder as the result of an infinite Doppler shift experienced by the inertial observer.

If we follow the standard definition and assume that the centrifugal force is an "apparent" force, appearing only in non-inertial frames of reference then one may wonder: is it possible to derive above-described dynamics of the bead inside the rotating pipe in the laboratory inertial frame (LF), relative to which the pipe rotates? Below it will be shown that all quantitative results of Ref. [6] are not related at all to peculiarities of the frame in which the problem is considered. We shall discuss also two other interesting aspects of the problem, viz. surprising and remarkable analogy of the "pipe+bead" *gedankenexperiment with*: (a) motion of classic simple gravity pendulum, and (b) the radial geodesic infall motion of a test particle in Schwarzschild geometry.

2.2 The original solution: LF study

In this section, we reconsider the *gedankenexperiment* in the LF, where the space–time is Minkowskian. Our purpose is to show that the peculiar dynamics of the system disclosed in the RF are not related at all to the arguable concept of a reversible or irreversible centrifugal force; on the contrary, it can be easily derived within the LF where no inertial forces are present.

Owing to the symmetry of the system, it is convenient to write the spatial part of the metric in polar coordinates:

$$ds^2 \equiv -d\tau^2 = -dt^2 + r^2 d\varphi^2 + dr^2, \tag{26}$$

The metric has three nonzero Christoffel symbols:

$$\Gamma^r_{\phi\phi} = -r, \quad \Gamma^\phi_{r\phi} = \Gamma^\phi_{\phi r} = 1/r. \tag{27}$$

The radial velocity and angular velocity of a bead/particle are defined as:

$$v \equiv dr/dt, \quad \omega \equiv d\phi/dt, \tag{28}$$

while the Lorentz factor is:

$$\gamma \equiv dt/d\tau = \left(1 - r^2\omega^2 - v^2\right)^{-1/2}, \tag{29}$$

The motion of the bead is free in the radial direction, while the pipe wall acts on the bead with tangential force of reaction. The radial component of the equation of motion

$$\frac{d^2 r}{d\tau^2} + \Gamma^r_{\alpha\beta}\left(\frac{dx^\alpha}{d\tau}\frac{dx^\beta}{d\tau}\right) = 0, \tag{30}$$

is reduced to:

$$\frac{d^2 r}{d\tau^2} = r\left(\frac{d\phi}{d\tau}\right)^2 = r\omega^2\gamma^2. \tag{31}$$

As far as $dr/d\tau = \gamma v$ and

$$\frac{d\gamma}{dt} = \gamma^3 v\left(\omega^2 r + \frac{dv}{dt}\right), \tag{32}$$

it is quite easy to calculate that:

$$\frac{d^2 r}{d\tau^2} = \gamma^2\left[\left(1 + \gamma^2 v^2\right)\frac{d^2 r}{dt^2} + \omega^2 r \gamma^2 v^2\right], \tag{33}$$

and it is equally straightforward to show that:

$$1 - \gamma^2 v^2 = \gamma^2\left(1 - r^2\omega^2 - 2v^2\right), \tag{34}$$

$$1 + \gamma^2 v^2 = \gamma^2\left(1 - r^2\omega^2\right), \tag{35}$$

which, in turn, after taking into account (31) and (33), allows us to derive the following equation:

$$d^2rdt^2 = \frac{\omega^2 r}{1-\omega^2 r^2}\left(1-\omega^2 r^2 - 2v^2\right). \tag{36}$$

One can easily see that it is exactly the same Eq. (15), which has been derived in the previous subsection within the RF analyses of the same problem!

It is worthwhile to note that taking into account (34) and (35), we can rewrite this equation in the following, surprisingly beautiful, form:

$$\frac{d^2r}{dt^2} = \left(\frac{1-\gamma^2 v^2}{1+\gamma^2 v^2}\right)\omega^2 r. \tag{37}$$

Apart from its elegant form, this equation turns out to be quite informative because it distinctly shows asymptotic peculiarities of the bead dynamics. Namely, if the motion is nonrelativistic ($\gamma v \ll 1$), (37) reduces to the usual classical equation for centrifugal acceleration: $d^2r/dt^2 = \omega^2 r$; while in the ultrarelativistic limit ($\gamma v \gg 1$), the sign of the right-hand side of (37) changes (!) to the opposite: $d^2r/dt^2 = -\omega^2 r$. Note, furthermore, that when $\gamma_0 v_0 = 1$ (viz. $v_0 = 1/\sqrt{2}$) above-mentioned remarkable acceleration sign reversal occurs from the very beginning of the motion!

Another mathematically elegant feature of the problem is that (36) can be rewritten as a differential equation for the function $y \equiv v^2(r)$. The resulting equation is:

$$\frac{d^2y}{dt^2} + \frac{4\omega^2 r}{1-\omega^2 r^2}y = 2\omega^2 r, \tag{38}$$

with the solution, which can be written as:

$$y(r) = \left(1-\omega^2 r^2\right) + const\cdot\left(1-\omega^2 r^2\right)^2. \tag{39}$$

In particular, for the "conventional" initial condition when initially ($t = 0$) the bead is situated on top of the pivot $r_0 = 0$ and has a velocity $v_0 \neq 0$, the latter equation implies that $const = v_0^2 - 1$. Therefore, for the function $v(r)$ we obtain the result that had been derived in Ref. [6]:

$$v(r) = \sqrt{\left(1-\omega^2 r^2\right)\left[1 - \frac{\left(1-\omega^2 r^2\right)}{\gamma_0^2}\right]}. \tag{40}$$

2.3 An interesting pendulum analogy

The *gedankenexperiment* reveals" surprising dynamics of relativistically rotating particles. But it happens to be remarkable also due to a couple of nontrivial and noteworthy analogies.

One of them shows when we introduce new set of variables:

$$\phi \equiv 2\,arccos\,(\omega r) \tag{41}$$

$$\lambda \equiv \omega t \tag{42}$$

$$\Omega^2 \equiv 1 - v_0^2 \tag{43}$$

and after, some uncomplicated calculations, find out that our equation of motion reduces to the following, remarkably familiar equation:

$$\frac{d^2\phi}{d\lambda^2} + \Omega^2 sin\phi = 0, \tag{44}$$

we arrive at the well-known *pendulum equation*!

Note that one could easily derive (44) from (43) by rewriting it in above-introduced notation as $d\phi/d\lambda = -2\sqrt{1 - \Omega^2 sin^2(\phi/2)}$, and taking one more derivative by λ. It is common knowledge that (44) describes nonlinear oscillations of a free "simple gravity" pendulum. In fact, it is closely related to another big achievement for Christiaan Huygens, who studied in his "Horologium Oscillatorium" in 1673. With this analogy, now being so apparent, the striking resemblance of Ref. [6] solutions with a pendulum motion becomes more transparent and understandable. If we introduce the concept of an *"analogous pendulum"*, governed by (44), then the initial conditions ($r_0 = 0$, $(dr/dt)_{t=0} = v_0$) are replaced by $\phi_0 = \pi$ and $(d\phi/d\lambda)_{\lambda=0} = -2v_0$. Therefore, it turns out that the analogous pendulum rotates in a vertical plane with the effective frequency Ω. Remarkably enough, the time interval, needed by the bead to reach $\omega r = 1$ ("light cylinder"), corresponds to the time needed by the analogous pendulum to reach its stable equilibrium ($\phi = 0$) point. This time interval is finite, as it, certainly should be.

2.4 A captivating black hole analogy

Another striking analogy turns out to be even more unexpected. One can show that the "bead-pipe" *gedankenexperiment* features an unusual analogy with a specific kind of geodesic motion in Schwarzschild geometry!

Let us consider a radial geodesic "fall" of a test particle onto a Schwarzschild black hole with M mass. Let a radial velocity of the particle at infinity be V_∞ pointed inwards. If one denotes by $E = \gamma_\infty \equiv \left(1 - {V_\infty}^2\right)^{-1/2} > 1$ the specific energy of the particle per its rest mass, then for its radial velocity relative to the observer at infinity $V_{\hat{r}} \equiv \sqrt{g_{rr}} dr/dt$ one gets

$${V_{\hat{r}}}^2 = \frac{E^2}{4} - \left[\frac{E}{2} - \frac{1}{E}\left(1 - \frac{2M}{r}\right)\right]^2, \tag{45}$$

(See, e.g., Ref. [23], where this equation is written in different notation). For the quantity $dV_{\hat{r}}/dt$ (called by McVittie "nontenzor radial acceleration of the particle") one gets:

$$\frac{dV_{\hat{r}}}{dt} = \frac{2M}{Er^2}\left[\frac{E}{2} - \frac{1}{E}\left(1 - \frac{2M}{r}\right)\right]\sqrt{1 - \frac{2M}{r}}. \tag{46}$$

Inspecting the expression it is easy to see that the acceleration of the particle is negative (that is, the modulus of the particle's infall velocity is increasing), until the particle reaches the radial distance $r_1 = 4M/(2 - {\gamma_\infty}^2)$, where the acceleration *changes* its sign, and $V_{\hat{r}}$ reaches its maximum velocity $V_{max} = \gamma_\infty/2$.

More precisely, we have the following regimes of the motion:

- $V_\infty \ll 1$ ($\gamma_\infty \approx 1$): the particle begins to move with increasing speed, at $r_1 \approx = 4M$ the speed reaches its maximum value ($V_{max} \approx 1/2$) and further on it decelerates, down to zero radial velocity, when the particle approaches the black hole horizon $r \to 2M$;

- $V_\infty = \sqrt{2}/2$, ($\gamma_\infty = \sqrt{2}$): the "threshold" case; with $dV_{\hat{r}}/dt = 0$ ($V_{max} = V_\infty$, $r_1 \to \infty$), and the particle decelerating smoothly, all along, down to the horizon;

- $V_\infty > \sqrt{2}/2$: the acceleration is positive during the whole course of the motion.

It is easy to notice that this scenario remarkably resembles (even quantitatively!) the one in Ref. [6] *gedankenexperiment*. This likeness happens due to the remarkable similarity between (45) and the corresponding equation for dr/dt from Ref. [6], which may be written as:

$$\left(\frac{dr}{dt}\right)^2 = \frac{E^2}{4} - \left[\frac{E}{2} - \frac{1}{E}\left(1 - \omega^2 r^2\right)\right]^2, \tag{47}$$

The resemblance is not incidental, of course, but is related to the likeness of the space–time (4) in the rotating pipe:

$$ds_p{}^2 = -\left(1 - \omega^2 r^2\right)dt^2 + dr^2, \tag{48}$$

with the Schwarzschild space–time along a radial geodesic ($\theta = const$, $\varphi = const$):

$$ds_s{}^2 = -(1 - 2M/r)dt^2 + \frac{dr^2}{1 - 2M/r}. \tag{49}$$

Comparing, for example, lapse functions for these two metrics ($\alpha_p \equiv \sqrt{-g_{tt}} = \sqrt{1 - \omega^2 r^2}$, and $\alpha_s \equiv \sqrt{-g_{tt}} = \sqrt{1 - 2M/r}$) one can see that α_p and α_s become infinite at the light cylinder and horizon, respectively. Obviously, the likeness is not complete: the spatial part of the rotating pipe metric is flat, while the spatial part of (49) metric is curved ($g_{rr} \neq 1$). The latter difference is also essential: it ensures finiteness of the time t^* [6], needed by the bead to reach the light cylinder $r^* \equiv \omega^{-1}$. As it is well known, an analogous time interval for a test particle to reach the Schwarzschild black hole horizon, as measured by the distant observer, is infinitely long!

2.5 A centrifugal force "reversal"?!

It is well known that a generalization of Newton's second law

$$\vec{F} = d\vec{p}/dt = d\left(m\vec{v}\right)/dt \tag{50}$$

is the most useful and convenient definition of a (three-) force $\vec{F}$ in special relativity. According to Rindler ([24], p. 88) *"This definition has no physical content until other properties of force are specified, and the suitability of the definition will depend on these other properties."*

For real applied forces, arising in relativistic dynamics, this definition is physically self-consistent and proved to be the most appropriate. But, how one should define *inertial* forces in relativity!? Miller and Abramowicz suggested [21] using the same general method for these forces as well. The approach is fully justified but we would like to emphasize one aspect of the problem.

The definition contains a quantity $m(t) = m_0\gamma(t)$, which has, clearly, the meaning of the relativistic mass in the LF. However, note that this quantity is used for the definition of another physical quantity—centrifugal force—which exists in another *non-inertial* frame of reference. A question arises: is it self-consistent to define a physical variable in one frame by means of another variable defined in another frame? Certainly, the time-variable mass may be defined also in RF, but it would not be equal to $m(t)$. The important point is that $\gamma(t)$, having in the LF the meaning of Lorentz factor, has not the same meaning in the RF because Lorentz factor of the moving particle is not invariant between frames [24].

This circumstance appears evident in the "$1 + 1$" formulation of the same problem. In fact, for the two-dimensional curved metric (4) in the RF $V \equiv v/\alpha_p$, and $\Gamma = \left[(1 - \omega^2 r^2)/(1 - \omega^2 r^2 - v^2)\right]^{1/2}$. Accordingly, it is possible to define relativistic mass in the RF as $M(t) \equiv m_0\Gamma$ and to write, instead of Eq. (50), the following definition:

$$F^* = \frac{d}{dt}(MV) = \frac{M\omega^2 r}{\alpha_p}. \tag{51}$$

This definition is already made by means of the true Lorentz factor of the bead as measured in the RF. Just like (50) it, also, gives "irreversible" centrifugal force but lacks the attractive simplicity of (50). However, we should always remember that the mass, which the RF observer actually *measures*, is $M(t)$ and is not $m(t)$. It seems, therefore, more consistent to express the physical quantity existing in the particular frame (centrifugal force f_c, which exists in the RF) through the other physical quantity $M(t)$ defined and measured in the same frame.

Despite this uncertainty, we should like to note that the importance of the mass variation effect, noticed by Miller and Abramowicz, is a very remarkable and important feature of this problem. As it appears, the capability of mass to vary drastically affects the dynamics of the motion. But, is it appropriate and logically justified to describe this secondary dynamical effect as the action of some "negative self-thrust" force!? Saying "secondary" we do not mean its significance but, rather, its status in the causal order of true physical reasons. If one introduces it that would be yet another "apparent" force, like the centrifugal force. Actual dependence of the bead mass on time is governed explicitly by the concrete kind of its motion in the LF, which is entirely determined by the outer *real* force, tangential pipe reaction force, applied to this moving body.

3. Astrophysical applications and conclusions

In 2003, the problem of motion of particles on relativistically rotating prescribed trajectories considered in Ref. [6] for a straight trajectory case (*gedankenexperiment*) was generalized. Rogava, Osmanov, and Dalakishvili considered the motion of rotationally driven particles along flat trajectories of arbitrarily curved shape [7]. In this paper, the problem was again studied on the level of the idealized *gedankenexperiment*, both in the laboratory (LF) and in the rotating (RF) frames of reference. For the simple example of the Archimedes spiral, it was found that the dynamics of these particles may involve both accelerative and decelerative modes of motion. Moreover, it was also found that there are special solutions, which remain force-free during the whole course of the motion.

Additionally, basic equations were outlined and the scheme for the solution of the more general version of the same problem was given, where the angular velocity of the rotating system is *not* assumed to be constant. Instead, it is required that the full triple system—rotator+pipe+bead—is conservative and the rotator is allowed to exchange perceptible portions of energy with the bead. The latter setup is a significant step closer to real astrophysical situations where, for instance, a rapidly rotating neutron star acts as a rotator and its magnetic field lines play the role of wires (pipes) along which particles are centrifugally accelerated.

One astrophysically important difference from the linear pipe case [6] is that for curved prescribed trajectories and the motion of the accelerated particle is not any more radially bounded: there exist regimes of motion when the bead may reach infinity! This result has a simple physical explanation. For the case of the linear pipe, rotating with the constant angular velocity, the natural limit of the radial motion is given by the light cylinder radius. However, in the case of the curved pipe, even when it rotates at the constant rate, the bead moves both in radial and azimuthal directions, following the curvature of the prescribed trajectory and having a variable angular velocity $\Omega(t)$. It means that now the role of the *"effective light cylinder* is played by the time-dependent quantity $R_L(t) = \Omega^{-1}(t)$, and all those radial distances for which $r(t) < R_L(t)$ become accessible! Therefore, if both $r(t)$ and $R_L(t)$ are monotonously increasing functions, but the former stays always smaller than the latter (e.g., that is the case for the Archimedes spiral) then the bead can reach infinity because wherever it is at any given moment of time, the light cylinder is still ahead of it and the centrifugally accelerated bead will never "overtake" the light cylinder!

4. Acceleration of particles by rotating magnetospheres in AGN

It is well-known that some blazars, such as Mrk 421, Mrk 501, PKS 2155–304, 1ES 2344–514, H1426 + 428, and 1ES 1959 + 650, emit ultra-high-energy, several TeV $(1\ \mathrm{TeV} \equiv 10^{12}\ \mathrm{eV})$ photons and constitute a special class of so-called *"TeV blazars"*. The standard model for a blazar assumes the presence of a supermassive black hole, surrounded by an accretion disc and ejecting twin relativistic jets, one of which is seen by a terrestrial observer almost end-on. Usually, broadband emission spectra of these objects contain two components: the low-energy (from radio to optical/UV) part associated with synchrotron radiation and the high-energy (X- and γ- rays) component formed by the inverse Compton scattering (ICS) of softer photons. The presence of the latter part of the spectrum is normally explained on the basis of the synchrotron self-Compton (SSC) model. However, the origin of accelerated electrons, the physical mechanism responsible for their efficient acceleration is a matter of uncertainty.

Proposed mechanisms such as the Fermi-type acceleration processes or repetitive acceleration of electron-positron pairs as a feedback mechanism could account for the observed high energy emission up to 20 TeV. However, the Fermi-type acceleration in relativistic jets is efficient only if the seed population of "pre-accelerated" electrons is present, possessing high enough ($\gamma_{\max} \geq 10^2$) Lorentz factors. The provenience of the "pre-acceleration" is not well-understood.

Since pioneering work by T. Gold [25, 26] it has been proposed that centrifugally driven outflows (CDOs) consisting of centrifugally accelerated particles may acquire extremely high energies. The idea has often been discussed [8, 27] in the context of pulsar emission theory. Moreover, regarding the AGNs, it has been shown that CDOs from accretion disks may occur if the poloidal magnetic field lines are inclined at an angle $\leq 60°$ to the equatorial plane of the disk [28].

Osmanov et al. [9] argued that the presence of the CDO would imply that despite the intense energy losses via UV radiation (when due to ICS, soft photons are upscattered on electrons and gain energy from them), electrons may still reach very high, $\gamma_{max} \sim 10^5$, Lorentz factors. It was surmised that if robust enough, this mechanism could account not only for the "pre-acceleration" of electrons but could be considered as an independent bona fide alternative mechanism for the generation of high-energy emission (up to $\sim$20 TeV) of TeV blazars!

Gangadhara and Lesch [29] examined the role of the centrifugal force on the dynamics of electrons moving along straight magnetic field lines, situated in the equatorial plane and corotating with the spinning AGN. They showed that an upscattering of low-energy photons against accelerated electrons may readily lead to the generation of the nonthermal x-ray and γ-ray emission. Later on, the problem was critically re-examined by Rieger and Mannheim [13], checking whether the rotational energy gain of charged particles, moving along straight magnetic field lines, is limited not only by ICS but also by the expected breakdown of "the-bead-on-the-wire" approximation. The latter is a synonymous term for the "bead+pipe" approximation used in our *gedankenexperiment*. The breakdown happens in the vicinity of the light cylinder, where a Coriolis force acting on the particle and trying to "tear it off" the field line would exceed the Lorentz force "binding" the particle to the field line.

Bearing in mind that in real, three-dimensional astrophysical situations (e.g., jets), the magnetic field lines are not localized in the equatorial plane but are inclined with respect to it. Osmanov et al. [9] examined the same problem considering the wide range of possible inclinations. It was shown that for a wide range of AGN, the mechanism responsible for limiting the maximum Lorentz factors is ICS, but under favorable conditions, it can allow particles to reach quite high Lorentz factors $\gamma_{max} \geq 10^5$. The breakdown of the-bead-on-the-wire approximation becomes important for the low luminosity ($< 10^{41}$ erg/s) AGN, when $\gamma_{max} \sim 10^8$. For higher luminosities ($> 10^{41}$ erg/s), it can be dominant only for relatively small inclination angles of the magnetic field lines with respect to the rotation axis ($\leq 10°$).

Therefore, it was argued [9] that CDOs could indeed be amply efficient to account for the TeV blazar emission. For the wide range of cases, ICS remains the dominant factor in limiting the maximum Lorentz factor of accelerated particles, letting them produce very high-energy photons [1–20] TeV. Obviously, an important restriction of the [9] approach is that only straight magnetic field lines were considered, whereas, in realistic astrophysical situations, the magnetic field lines are curved. It might be especially important if one considers particle dynamics on a longer time scale and over larger length scales when the curvature of the field lines cannot be neglected. As a matter of fact, the mathematical formalism for such a study is described in [7].

4.1 Centrifugal acceleration and gamma flares in Crab nebula

The Crab nebula is a source of almost steady high-energy emissions. Observations made from orbital probes (Fermi, SWIFT, and RXTE) showed evidence of its variability in the X\x-ray range. Several years ago the Fermi and AGILE satellites detected dazzling, brief, and strong bursts of gamma radiation above 100 MeV, with its source located in the Crab Nebula. Since then, several gamma-ray flares have been reported.

The most obvious source of radiation power in the nebula is the rotational energy damping $dE/dt = I\Omega d\Omega/dt = 5 \cdot 10^{38} erg/sec$ of the pulsar PSR 0532. Machabeli et al. [30] demonstrated that via the centrifugal acceleration mechanism, it is possible to pump energy from the neutron star's vast rotational kinetic energy storage, $\sim 5 \times 10^{38}$ erg s − 1, to proper electrostatic plasma (Langmuir) oscillations. Furthermore, it was shown that the growth rate of the perturbations is maximum in the "superluminal" area, where the phase velocity of perturbations exceeds the speed of light. That is why in this region, the condensate of plasmons is formed, which is transferred to the Crab Nebula. The transfer of the energy of the plasmon condensate from the pulsar magnetosphere to the nebula over a huge distance, $3 \times 10^{17} cm$, takes place practically without any tangible losses.

But in the nebula, unlike the pulsar magnetosphere, apart from electrons and positrons, there are also protons. That is why a *modulation instability* is developed, which leads to the collapse of Langmuir waves; viz. a cavity is formed, which collapses, and on the final stage of the collapse, involved particles attain very high Lorentz factors, resulting in the powerful synchrotron emission of the nebula. The collapse stops at the scale of a few Debye radii.

It was also shown that in the course of the active phase of the collapse in the cavity due to the influence of nonlinear processes on the polarization properties of the medium, *self-trapping* [31] of the synchrotron radiation (generated within the cavity) takes place. If the conditions for the appearance of the self-trapping are fulfilled for certain values of emitted wave frequencies, for others, both higher and lower values of the frequency, they are not satisfied. It means that giant bursts of energy release have to happen in narrow energy (frequency) ranges. Therefore, it would naturally explain abrupt, burst-like increases of the radiation intensity within narrow bands, impressively explained by self-trapping in the framework of nonlinear optics (so-called "Askaryan effect" [32]). Waves propagating in the non-parallel direction to the line of sight are bent and directed to the observer due to self-trapping. Ultimately it leads to the required increase of the radiation intensity and, at the same time, unlike other mechanisms, it does not require any additional sources of energy.

4.2 Self-trapping as a beaming mechanism for Fast Radio Bursts!?

Bearing in mind what we have just said about self-trapping, it is most reasonable to assume [33] that the principal reason for the beaming of electromagnetic radiation leading to Fast Radio Bursts (FRBs) could be the nonlinear self-trapping phenomenon. This mechanism implies that the part of the radiation beam directed toward the observer is augmented by its outer part, which in the absence of the self-trapping would not be focused toward the same point. As a result, the observer, while the beam is being self-trapped, sees an enhanced intensity of radiation in a very narrow frequency range. This scenario is robust and fully autonomous because unlike many other mechanisms, it does not require additional external sources of energy. Besides, self-trapping depends on quite a large number of parameters, in particular, on a specific proper value of the ratio of the wave amplitude to the amplitude of an incident electrostatic wave, on the temperature of the medium, and on the direction of these waves relative to the line of sight. Any, even the slightest, deviation of any of these parameters from 'favorable' values may lead to the disappearance of an observable radiation burst. This is why this is an extremely finely tuned, random, and rarely occurring event. This circumstance guarantees that an FRB event, if indeed being caused by the arrival of the self-trapped self-focused enhanced beam to the observer, is a totally random and extremely rare natural phenomenon.

If our model is correct and relevant to actual FRBs, the self-trapping condition may hold in a given direction only for a very short interval of time. Hence, it is logical to surmise that the probability of the coincidence between the line of sight and the direction of self-trapping has to be quite small. Paradoxically enough, what could be a serious drawback for a commonly occurring phenomenon that in this case 'works' just in the opposite way, it strengthens our confidence that self-trapping could be an important physical factor contributing to the appearance of this extremely rare and energetic phenomenon—fast radio bursts or FRBs.

Obviously, the self-trapping mechanism does not exclude other physically plausible, repetitive or non-repetitive and catastrophic or non-catastrophic mechanisms proposed for FRBs. We tend to believe that self-trapping may be one of a number of very efficient 'beaming' mechanisms that might be needed for interpreting FRBs as narrowly beamed radio bursts [34]. We suppose that the prime reason for any single burst appearance could be related to its generic source, located in a distant galaxy, and causing a giant outburst in the radio range. The mechanism of this primary outburst may be related to one of the previously suggested plasma mechanisms.

Presumably, these outbursts might be observed without any additional amplification if they occur on intergalactic distance scales. That was Machabeli et al. guessed in Ref. [33]. Shortly afterward first-ever FRB from within our galaxy was detected [35], confirming our expectation since it was associated with a rapidly rotating neutron star, magnetar SGR 1935 + 2154 about 30,000 light-years away in the Vulpecula constellation. The STARE2 team independently observed the burst [36], detected its fluence, and confirmed the connection between this burst (FRB 200428) and FRBs at extragalactic distances. Needless to say, a magnetar is a plausible candidate for the centrifugally driven relativistic acceleration of electrons with subsequent Langmuir collapse and self-trapping as the beaming mechanism. It only strengthened our expectation that FRB could be akin to, for instance, giant radio pulses occasionally observed from pulsars within our own Galaxy [33]. Actually, for intergalactic giant pulses, the beaming mechanism may not be even necessary for the burst to be detectable. But in order to make FRBs visible on extragalactic distance scales, some powerful additional beaming/amplification is required and this is where we believe self-trapping could play its important role.

With improved capabilities of several wide-field, broad-band surveys that recently became operational the observational situation may change dramatically. It is expected that the FRB field will change from the presently available small-number statistics up to hundreds or even thousands [37] of new FRBs per year. Therefore, in forthcoming years, one might expect that the number of detected FRBs will increase by orders of magnitude, and we could be able to see primary outbursts leading to FRBs without a need for their additional 'self-trapping-related' or any other kind of amplification. But within this large number of events, there will be a relatively small fraction of peculiar bursts undergoing secondary self-trapping amplification and having distinctly peaked structures.

5. Conclusions and future prospects

Above given three examples of supposed astrophysical appearance were meant to emphasize that centrifugal acceleration may certainly be one of most efficient converters of rotational energy into the energy of directed motion and may lead to the generation of powerful observational signatures. The universe is full of rotating plasma flows containing enormous amounts of kinetic energy. There are a few physical mechanisms able to transform a considerable part of the mechanical energy into radiation energy. For instance, so-called "nonmodal" processes [38, 39], taking place in flows with nonuniform velocity fields (shear flows), are capable to infuse a part of the regular rotation-related energy in waves and instabilities appearing in these flows [40, 41]. However, none of these nonmodal mechanisms has efficiency even remotely comparable with the efficiency of the centrifugal acceleration mechanism.

Obviously, centrifugal acceleration has to be of particular importance in objects with ultra-strong magnetic fields: pulsars, magnetars, AGN, etc. In these astronomical situations, magnetic field lines provide proper "prescribed trajectories" for relativistic particles and can accelerate them up to very high Lorentz factors. That is why the issue of centrifugal acceleration occupies an important role in the domain of relativistic astrophysics.

Theoretically speaking the most principal, crucial challenge is to move from highly idealized, "gedankenexperiment" setups, and various "toy" problems up to more realistic models, replacing mechanical "bead-on-wire" approximation with real, strong magnetic fields serving as genuine accelerators of relativistic particles in related astronomical objects. We could easily envisage that with the development of observational capabilities in relativistic astrophysics possible areas of applications for the centrifugal acceleration mechanism in high-energy astrophysics will widen further. Obviously, a number of intriguing problems remain to be understood. For instance, how centrifugal acceleration operates in conjunction with space–time "rotation" nearby black holes!? In Refs. [10, 20], a theoretical formalism was developed allowing to study of centrifugally induced phenomena in general static and isotropic metrics, associated with rotating relativistic bodies. One can imagine, for example, that in a Kerr black hole ergosphere, the interlace of centrifugal acceleration with Penrose process [42] may lead to interesting physical peculiarities, some of which may turn out to be important in the astronomical-observational context.

Acknowledgements

Andria Rogava's research was supported by Shota Rustaveli National Science Foundation of Georgia (SRNSFG) grant FR-18-14747.

Author details

Andria Rogava
Institute of Theoretical Physics, Ilia State University, Tbilisi, Georgia

*Address all correspondence to: andria.rogava@iliauni.edu.ge

References

[1] Yoder J. Christiaan Huygens' Great Treasure. Trattrix. 1994;**3**:1

[2] Abramowicz MA. Centrifugal Force - a Few Surprises. Monthly Notices of the Royal Astronomical Society. 1990; **245**:733

[3] Abramowicz MA, Carter B, Lasota JP. Optical Reference Geometry for Stationary and Static Dynamics. General Relativity and Gravitation. 1988;**20**:1173

[4] Abramowicz MA, Prasanna AR. Centrifugal Force Reversal Near a Schwarzschild Black-Hole. Monthly Notices of the Royal Astronomical Society. 1990;**245**:720

[5] Abramowicz MA, Miller JC. Ellipticity Behaviour of Relativistic Maclaurin Spheroids. Monthly Notices of the Royal Astronomical Society. 1990;**f245**:729

[6] Machabeli GZ, Rogava AD. Centrifugal force: A Gedanken Experiment. Physical Review A. 1994; **50**:98

[7] Rogava AD, Dalakishvili G, Osmanov ZN. Centrifugally Driven Relativistic Dynamics on Curved Trajectories. General Relativity and Gravitation. 2003;**35**:1133

[8] Gangadhara RT. Emission in Spin-powered Pulsars and Polarization Position Angle. Astronomy & Astrophysics. 1996;**314**:853

[9] Osmanov ZN, Rogava AD, Bodo G. On the Efficiency of Particle Acceleration by Rotating Magnetospheres in AGN. Astronomy & Astrophysics. 2007;**470**:395

[10] Gudavadze I, Osmanov ZN, Rogava AD. On the Role of Rotation in the Outflows of the Crab Pulsar. International Journal of Modern Physics D. 2015;**24**:1550042

[11] Machabeli GZ, Rogava AD, Chkheidze N, Osmanov ZN, Shapakidze D. The Crab Nebula Energy Origin and its High Frequency Radiation Spectra. Journal of Plasma Physics. 2016;**82**: 635820305

[12] Osmanov ZN, Rieger F. Pulsed VHE Emission from the Crab Pulsar in the Context of Magnetocentrifugal Particle Acceleration. Monthly Notices of the Royal Astronomical Society. 2017;**464**: 1347

[13] Rieger FM, Mannheim K. Particle Acceleration by Rotating Magnetospheres in Active Galactic. Astronomy & Astrophysics. 2000;**353**: 473

[14] Dalakishvili GT, Rogava AD, Berezhiani VI. Role of Radiation Reaction Forces in the Dynamics of Centrifugally Accelerated Particles. Physical Review D. 2007;**78**:045003

[15] Bakhtadze GG, Osmanov Z, Berezhiani VI. Centrifugal Acceleration in the Isotropic Photon Field. International Journal of Modern Physics D. 2017;**26**:1750135

[16] Arsenadze G, Osmanov Z. Particles on the Rotating Channels in the Wormhole Metrics. International Journal of Modern Physics D. 2017;**26**:1750153

[17] Shapiro SL, Teukolsky SA. Black Holes, White Dwarfs and Neutron Stars. New York: Wiley; 1983

[18] Landau LD, Lifshitz EM. Mechanics. Third edition. Oxford: Pergamon; 1976

[19] Abramowitz M, Stegun IA. Handbook of Mathematical Functions. National Bureau of Standards; 1964

[20] Khomeriki G, Rogava AD. Centrifugally Driven Relativistic Particles: General Treatment and New Solutions. General Relativity and Gravitation. 2020;**52**:1

[21] Miller JC, Abramowicz MA. Comment on "Centrifugal force: A gedanken experiment". SISSA ref. 178/94/A. 1994

[22] de Felice F. Dynamics on a rotating disk. Physical Review A. 1995;**52**:3452

[23] McVittie GC. General Relativity and Cosmology. London: Chapman and Hall; 1956

[24] Rindler W. Essential Relativity. New-York: Springer-Verlag; 1980

[25] Gold T. Rotating Neutron Stars as the Origin of the Pulsating Radio Sources. Nature. 1968;**218**:731

[26] Gold T. Rotating Neutron Stars and the Nature of Pulsars. Nature. 1968;**221**:25

[27] Contopoulos I, Kazanas D, Fendt C. The Axisymmetric Pulsar Magnetosphere. Astrophysical Journal. 1999;**511**:351

[28] Blandford RD, Payne DG. Hydromagnetic Flows From Accretion Disks and the Production of Radio Jets. Monthly Notices of the Royal Astronomical Society. 1982;**199**:883

[29] Gangadhara RT, Lesch H. On the Nonthermal Emission in Active Galactic Nuclei. Astronomy & Astrophysics. 1997;**323**:L45

[30] Machabeli GZ, Rogava AD, Shapakidze D. On the Origin and Physics of Gamma Flares in Crab Nebula. Astrophysical Journal. 2015;**814**:38

[31] Chiao R, Garmire E, Townes C. Self-Trapping of Optical Beams. Physical Review Letters. 1964;**13**: 479

[32] Askaryan GA. Interaction Between Laser Radiation and Oscillating Surfaces. Journal of Experimental and Theoretical Physics. 1962;**15**:116

[33] Machabeli GZ, Rogava AD, Tevdorashvili B. Self-trapping as the possible beaming mechanism for FRBs. Monthly Notices of the Royal Astronomical Society. 2019;**489**:5688

[34] Katz J. Are Fast Radio Bursts Wandering Narrow Beams? Monthly Notices of the Royal Astronomical Society. 2017;**467**:L96

[35] Zhang SN. et al. Insight-HXMT Detection of a Bright Short x-ray Counterpart of the Fast Radio Burst from SGR+2154. 2020;**13687**:1

[36] Bochenek CD et al. STARE2: Detecting Fast Radio Bursts in the Milky Way. Publications of the Astronomical Society of the Pacific. 2020;**132**:034202

[37] Connor L. Interpreting the Distributions of FRB Observables. Monthly Notices of the Royal Astronomical Society. 2019;**487**:5753

[38] Rogava AD, Mahajan SM, Bodo G, Massaglia S. Swirling Astrophysical Flows – Efficient Amplifiers of Alfvén Waves!? Astronomy & Astrophysics. 2003;**399**:421

[39] Rogava AD, Poedts S, Heirman S. Are Galactic Magnetohydrodynamic Waves Coupled? Monthly Notices of the Royal Astronomical Society. 1999;**307**: L31

[40] Rogava AD, Osmanov Z, Poedts S. Self-heating and its Possible Relationship to Chromospheric Heating in Slowly Rotating Stars. Monthly Notices of the Royal Astronomical Society. 2010;**404**: 224

[41] Mahajan SM, Machabeli GZ, Rogava AD. Escaping Radio Emission from Pulsars: Possible Role of Velocity Shear. Astrophysical Journal. 1997;**479**:L129

[42] Penrose R, Floyd RM. Extraction of Rotational Energy from a Black Hole. Nature. 1971;**229**:177

Chapter 5

Time-Dependent Photoluminescence and Photoluminescence Excitation in Exciton Systems and Related Phenomena

John W. Kenney III, Joshua Jacobsen, Amanda Renfro, Isaac Muñoz and Ruth Christian

Abstract

The term "exciton" covers an extremely diverse range of materials, phenomena, processes, interactions, and experimental techniques. This review provides a general introduction-with selected descriptive examples-of excitonic systems with an emphasis on excitonic photoluminescence and photoexcitation spectroscopy in the ultrafast time-resolved femtosecond time domain.

Keywords: exciton, Frenkel, Wannier-Mott, hole, electron, time-resolved, femtosecond, ultrafast, luminescence, photoluminescence, graphene, graphene quantum dot, single-walled carbon nanotube, quasiparticle, 2D material, Bose-Einstein condensate

1. Introduction

The simplest example of an exciton is that of an electron (e^-) attracted to a hole (h^+) via the Coulomb force. The electron charge and hole charge are equal in magnitude and opposite in sign

$$|e^-| = |h^+| = e \tag{1}$$

where $e = 1.602 \times 10^{-19}$ C is the fundamental unit of electrical charge. The magnitude of the Coulomb force acting between the electron and the hole is given by the expression

$$F = \frac{e^2}{4\pi\varepsilon_0 r^2} \tag{2}$$

where r is the distance of separation between the electron and the hole and $\varepsilon_0 = 8.854 \times 10^{-12}$ $C^2/N{\cdot}m^2$ is the vacuum permittivity of free space. This electron–

hole unit is termed a quasiparticle. Excitons may be produced by bombarding the material of choice with photons whose energy $h\nu$ is greater than the energy band gap between the between the valence and conduction bands of the medium

$$h\nu > E_c - E_v = E_g \tag{3}$$

If this condition is met, negatively charged electrons are displaced within the medium and positively charged holes are formed in the medium, leaving the crystal lattice or material electrically neutral overall. A positively charged electron hole is thus created if the condition described in Eq. (3) is also met. The positive electron hole in the conduction band has diminished affinity for this localized hole due to repulsive Coulomb forces from large numbers of electrons that surround the hole and excited electron. Also, in a medium, Coulomb's law of force is modified by adding in a dimensionless dielectric constant term K to the Coulomb's Law expression given in Eq. (2).

$$F = \frac{e^2}{4\pi\varepsilon_0 K r^2} \tag{4}$$

In a vacuum, $K = 1.000$, but in a medium, $K > 1.000$. Thus, the attractive Coulomb force that acts between the electron and hole in an exciton is considerably less than the attractive Coulomb force that acts between the "orbiting" electron and the proton nucleus in a hydrogen atom.

The symmetry of the wavefunctions of an excitonic quantum system is virtually the same as the symmetry of the corresponding wavefunctions of the quantum system of atomic hydrogen. In atomic hydrogen, however, the electron e^- orbits around a positively charged nucleus, a proton p^+. In contrast, in an exciton, the electron e^- orbits around a positive hole h^+. The associated hydrogenic excitonic wavefunction exhibits symmetry properties quite similar to the symmetry properties exhibited by atomic hydrogen excitonic states (e.g., S, P, D ...).

Frenkel (in 1931) first proposed the idea of excitons as a means by which to describe the excitations of atoms (e.g., Na, K, Cl) embedded in a lattice of insulators (e.g., diamond, NaCl, $CaCO_3$). He further proposed that this excited atom-insulator system would be capable of supporting intra-lattice particle-like movements without net charge transfer. Excitons are typically treated in terms of two limiting cases of the dielectric constant K expression in Eq. (4):

Small K: Frenkel (F) Excitons Large K: Wannier-Mott (WM) Excitons.

The electron and hole in an exciton may each exhibit parallel or anti-parallel spins. The exchange interaction couples the spins, which yields exciton fine structure. An exciton may also show momentum dependence (i.e., k-vector dependence) in periodic lattices [1].

1.1 Frenkel excitons

The electron–hole Coulomb attraction may be strong, and its excitons will tend to be small (e.g., size of a unit cell) in a F exciton (**Figure 1**). It is even possible for molecular level excitons to be confined to a single molecule, as is often the case in fullerenes. Typical binding energies are on the order of 0.1 eV to 1.0 eV. F excitons occur frequently in excited d-d transition metal complexes and related systems with partially filled d-orbitals, such as a transition metal cation embedded in an insulator or

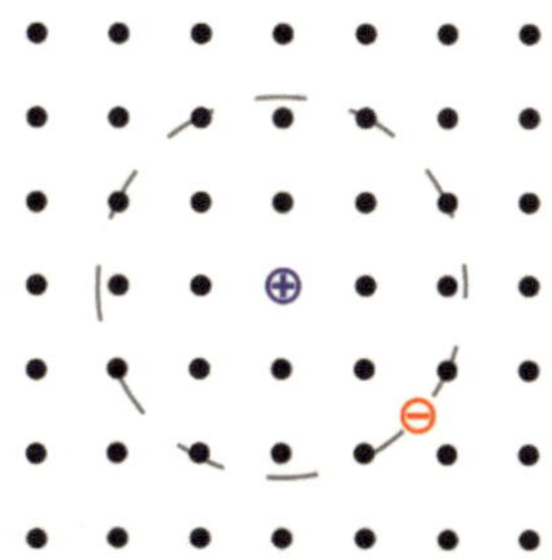

Figure 1.
F exciton in a lattice, showing the displaced electron in red and the hole in blue. The distance between the electron and hole in Eq. (2) is r. Image from public domain.

semiconductor matrix. While *d-d* transitions are forbidden by symmetry (i.e., all electronic states are g = gerade in the group O_h to which a ML_6 transition metal complex with six identical monodentate ligands belongs), these transitions can become weakly allowed by a lower symmetry vibration (u = ungerade), a crystal defect, or a lattice relaxation. Photon absorption by molecular organic crystals comprised of aromatic molecules (e.g., anthracene, tetracene, 1,0-phenanthroline, 2,2′-bipyridine) can give rise to the creation of a F exciton. Alkali halide crystals are also excellent substances from which F excitons may be generated [1].

1.2 Wannier-Mott excitons

In contrast to F excitons, which are characterized by small dielectric constants, the dielectric constants of WM excitons in many semiconductor systems can be quite large.

$$K_{\mathrm{WM}} > K_{\mathrm{F}}. \tag{5}$$

This results in a reduction of the Coulomb force between the electron and the hole and a consequent increase in the exciton radius in WM systems, such that the exciton radius may well become larger than the lattice spacing. The low effective mass of the electrons in a WM exciton is typical of semiconductors. For example, in the semiconductor gallium arsenide, GaAs, the relative permittivity is K = 12.8, the electron mass is $m_e = 0.067m_o$, and the hole mass is $m_h = 0.2m_o$ where the electron rest mass is $m_o = 9.109 \times 10^{-31}$ kg. WM excitons are primarily found in semiconductor crystals. These crystals are characterized by a small conduction-valence band energy gap ΔE_{CV} and high dielectric constant K. However, some WM excitons have been observed in liquids such as liquid Xe. In the literature, these excitons may also be referred to as *large excitons* [1].

1.3 Mixed WM and F excitons

A mixture of WM and F character may be observed in those excitons produced in single wall carbon nanotubes (SWCNTs). This situation arises because the dielectric function K of the nanotube is sufficiently large to allow the spatial aspect of the wavefunction to extend over one to several nm along the tube axis. If the screening in the dielectric or vacuum environment outside the nanotube is poor, large binding energies (0.4 eV to 1.0 eV) may be observed [1].

2. Ultrafast time-dependent photoluminescence and excitation spectroscopy of excitons: introduction

The classic spectroscopic probes of exciton structure and dynamics in many-body systems are steady-state PL spectroscopy, steady-state PL excitation spectroscopy, and steady state excitation spectroscopy, all ranging from the ultraviolet (UV) to the terahertz (THz). Raman and infrared spectroscopy also turn out to be quite useful in the study of excitons. More recently, many transient PL studies have been reported in the literature. Especially notable are optical-pump-THz-probe excitations and emissions facilitated by the rapid development of affordable and robust transient THz spectroscopic technology. For example, transient excitation (TE) spectroscopy was employed to measure the time-dependent excitation spectra of individual aligned SWCNTs previously selected and oriented by resonance Raman spectroscopic techniques on metallic SWCNTs (see [2]). Resonance Raman, atomic force microscopy, and TE were employed in concert on the same nanotube. TE ushers in the possibility of supplying both spectroscopic and dynamic information on individual SWCNT systems. These types of time-dependent, multi-spectroscopic measurements on precision-aligned samples provide the basis for exciting new opportunities that promise to unravel energy relaxation and energy migration pathways of excitons in metallic SWCNTs and elucidate the role of the substrate in the spectroscopy and dynamics of these fascinating materials. This transient femtosecond TE technique is especially useful when the sample is non-luminescent or only weakly luminescent and PL spectroscopy is not a viable option, which is often the case with weakly emitting metallic SWCNTs [3].

3. Excitons and angle resolved photoemission spectroscopy (ARPES)

Angle resolved photoemission spectroscopy (ARPES) is based on a straightforward application of Einstein's photoelectric effect. ARPES is a "photon in electron out" technique in which an incoming photon collides with a medium, ejecting an electron. Information about the medium under interrogation by ARPES is extracted from combined measurements of photon energy and the energy, momentum, and scattering angle of the ejected electron. ARPES is a valuable technique for the study of excitons. A beam of white light is generated in a beam line undulator and subsequently passed through a tunable grating monochromator to block all output photons except for those of wavelength λ needed for a specific ARPES specific experiment [4].

3.1 Excitons and angle time-resolved photoemission spectroscopy (tr-ARPES)

The impressive capabilities of the conventional time-independent, ARPES experiment to elucidate the nature and behavior of exciton systems can be further extended by the addition of time-resolved capabilities to the widely deployed angle-resolved capabilities. This addition yields essentially simultaneous time- and angle-resolved ARPES (tr-ARPES). Stroboscopic time resolution extends down to the ultrafast femtosecond (10^{-15} s) domain, providing high resolution temporal snapshots of exciton dynamics [5].

4. Linear and circular polarizations of exciton photoluminescence

Measuring the linearly polarized and circularly polarized PL spectra of an excitonic system can yield valuable information about the nature of the exciton from which the emission emerges. Simultaneous measurement of exciton PL spectra as a function of time, temperature, polarization, and pressure can provide exciting new insights regarding the nature of excitonic systems. Dynamic polarization measurements now exist that rely on an optical photoelastic modulator and a 1/4 wave plate to extract polarization-dependent spectral features. For example, circular polarized luminescence has been used to probe exciton coherence in disordered helical aggregates [6].

5. Temperature-dependence of excitons

The temperature dependence of exciton systems arises from several causes: PL intensity shifting, wavelength shifting, and exciton lifetime. The first major change is typically a decrease in PL intensity as the temperature is raised. Also, as temperature increases, there is a slight redshift of the peak wavelength. Finally, as temperature decreases, the lifetime of the exciton increases.

5.1 Changing photoluminescence intensity (Arrhenius formula)

$$I(T) = \frac{I_0}{1 + Ae^{\frac{-E_a}{k_B T}}} \tag{6}$$

The Arrhenius formula (Eq. 6) connects intensity, $I(T)$, of the PL at a given temperature for a specific material to the temperature T. I_0 is the PL intensity at 0 Kelvin (K), E_a is the activation energy due to thermal quenching, A is a constant, and k_B is the Boltzmann constant. This formula shows that as temperature decreases, the PL intensity of the exciton increases [7].

5.2 Temperature dependence of PL wavelength (Varshni formula)

The Varshni formula describes the relationship between energy band gap E_g and temperature T.

$$E_g(T) = E_g(0) - \frac{\alpha T^2}{\beta + T} \tag{7}$$

The graph of Eq. (7) (**Figure 2**) shows the correlation between E_g and T. Recall that $E_g(0)$ is the gap in energy between the bound state and the free state, α is the temperature coefficient, and β is the approximation to the Debye temperature. Note that the equation is only for materials that have a non-zero band gap, as is seen in some semiconductors. This shows that as temperature increases, the band gap becomes less restrictive, allowing electrons to pass through at a lower energy, causing the PL band to red-shift. The reason for this is phonon-exciton interactions [7].

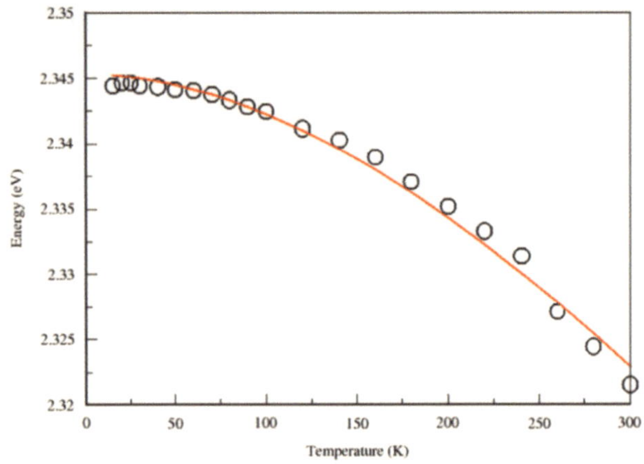

Figure 2.
Energy band gap (eV) with respect to temperature. Reprinted with permission from reference [8].

5.3 Exciton lifetime

The lifetime of an exciton is determined by its size and the temperature. An increase in temperature will increase the decay rate regardless of the size. A larger exciton will have a faster decay rate than a smaller exciton at the same temperature. A difference in size means a difference in surface-to-volume ratio; as the diameter decreases linearly, the volume decreases exponentially. However, because of the exponential decrease in volume, this increases the surface area to volume ratio, resulting in a slower decay rate for smaller diameter excitons. Regardless of size, as the temperature increases, the lifetime of an exciton decreases [7] (**Figure 3**).

5.4 Phonon-exciton interactions

The total PL emission width is describable as the sum of an inhomogeneous broadening parameter (Γ_{inh}) and several parameters representing homogeneous broadening arising from acoustic (Γ_{AC}) and optical (Γ_{LO}) phonon-exciton interactions

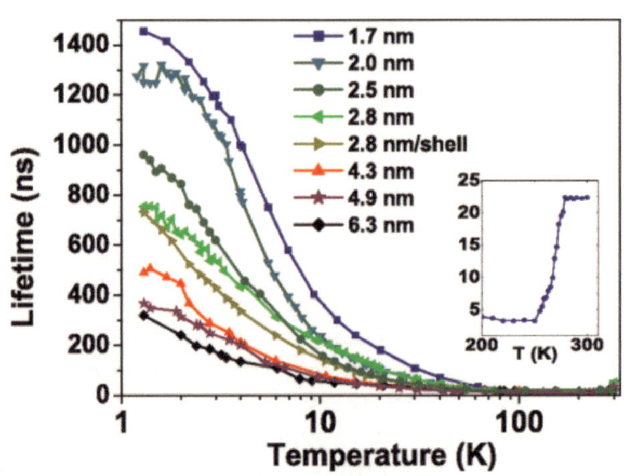

Figure 3.
Lifetimes of different sizes of excitons with respect to temperature. Permission requested from reference [9].

$$\Gamma(T) = \Gamma_{inh} + \Gamma_{AC}T + \Gamma_{LO}\left(e^{\frac{E_{LO}}{k_B T}} - 1\right)^{-1} \tag{8}$$

where $\Gamma(T)$ is the overall broadening parameter at temperature T, and E_{LO} is the longitudinal optical phonon energy [10].

6. Pressure modulation of exciton PL in the ultrafast femtosecond time domain

Recently, ultrafast femtosecond spectroscopic techniques have been adapted to produce pressure modulation of structure and exciton kinetics in the $1 \sim 15.45$ GPa range in two-dimensional (2D) halide perovskites (see [11]). A time-resolved ultrafast femtosecond PL investigation of excitonic systems as a function of pressure at low temperature has also been reported, along with pressure-tuned photon emission of trions and excitons in monolayer $MoSe_2$ in a diamond anvil cell (DAC) [12]. Work has also been done regarding pressure-induced ultrafast spectroscopic dynamics study of excitons in stretch-oriented poly(p-phenylenevinylene) (PPV) (see [13]).

7. Graphene and low-dimensional materials

Over the past two decades, graphene has emerged as one of the most interesting materials in carbon science, due to its chemical and thermal stability, mechanical flexibility, and high electron mobility [14]. Graphene is very similar to, and in fact a derivative of, the more common material, graphite (**Figure 4**). Graphene is a 2D sheet of carbon atoms arranged in a hexagonal honeycomb pattern, and graphite is the combination of many of these graphene sheets stacked on top of each other [15, 16]. Decades before this structure was ever realized in the real world, theoreticians coined the term "graphene" for these then imaginary 2D sheets. It was 17 years later, in 2004, that researchers Geim and Novoslov first synthesized the material at Manchester University, a feat for which they were awarded the Nobel Prize in Physics in 2010 [15, 17]. The low-dimensional nature of graphene, carbon nanotubes, and graphene

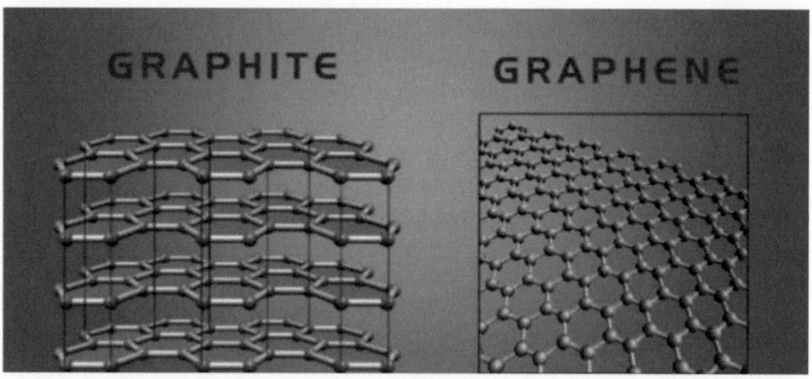

Figure 4.
A structural comparison of graphite and graphene. Reprinted with permission from Techinstro.

quantum dots (GQDs) gives rise to unusual optical phenomena that can be interrogated using PL spectroscopy [18].

7.1 Two-dimensional graphene

Femtosecond laser irradiation of graphene has produced unique spectra that have significant blue-shifted components, which is noteworthy due to the contrast between graphene's spectra and the spectra of traditional materials, whose spectra are normally dominated by their red-shifted components. This is due to the energy range that excitons formed by the irradiation populate. The photon collision initially creates excitons in graphene's conduction band at an energy E_0 of

$$E_0 = \frac{\hbar\omega_0}{2} = \frac{h\nu_0}{2} \tag{9}$$

where ω_0 is the angular frequency of the photon and ν_0 is the photon frequency. The excitons are then able to disperse throughout an energy of level from 0 to $2E_0$, due to graphene's zero-band-gap. Exciton recombination in the in the range from E_0 to $2E_0$ then produces blue-shifted photons. The 2D nature of graphene is the source of its zero-band-gap, and thus is the source of its strange PL spectra [18].

7.2 One-dimensional carbon nanotubes

There are three types of carbon nanotubes (CNTs): SWCNTs that are made up of one graphene wall, double-walled carbon nanotubes (DWCNTs) that consist of two graphene walls with one on the outside of the other, and multiwalled carbon nanotubes (MWCNTs), which have three or more graphene walls [19] stacked together concentrically (**Figure 5**). SWCNTs are quasi-one-dimensional [20] and can vary in structure; their geometry is determined by the chiral indices (n, m) [21] which specify the perimeter vector (chiral vector), of the CNT. The perimeter vector of a SWCNT is defined by

$$\hat{A} = (n, m) = n\hat{a}_1 + m\hat{a}_2 \tag{10}$$

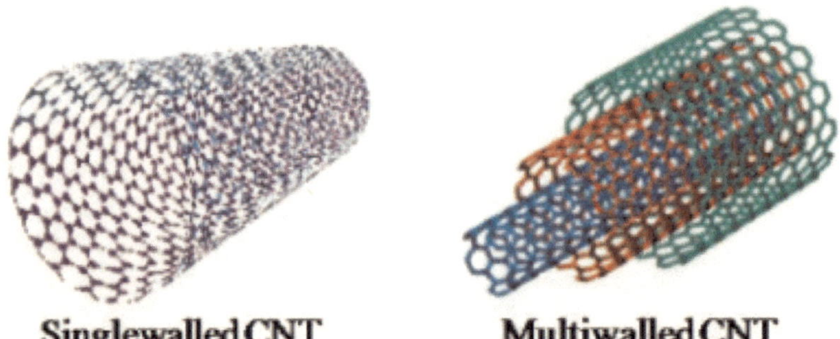

Figure 5.
Structure of a SWCNT and how multiple nanotube layers fit together to form a MWCNT. Reprinted with permission from cheaptubes.com.

where $\hat{A}$, is the perimeter vector and $(n\ m)$ are the chiral indices, and $\hat{a}_1$ and $\hat{a}_2$ are basis vectors. The magnitude A is

$$A = a_0\left(n^2 + m^2 + nm\right)^{1/2} \tag{11}$$

where a_0 is a basis vector of the graphene net and a_0 = 0.246 nm [22].

SWCNT PL originates in the lowest-energy band edge exciton state, but by obtaining and observing SWCNTs of increasing diameter, there is a possibility to tune PL to higher wavelengths [21]. Doing so results in an overall better understanding of SWCNTs. The diameter d of a SWCNT is [22]

$$d = \frac{A}{\pi} - \frac{a_0}{\pi}\sqrt{n^2 + m^2 + nm} \tag{12}$$

The challenge in tuning PL to higher frequency is finding a red-shifted wavelength that works, since there is a broad diversity of nanotube structures generated in typical synthesis procedures [21]. This is because SWCNTs are highly sensitive to the coupling with their environment which can have a dramatic impact on their electronic and optical properties. One of the most sensitive explorations of such environment-induced effects is the luminescence of semiconducting SWCNTs [23]. The structure variance is important to note because some structures are easier to study and there have been advancements made for selective growth of some structures but not others.

Recent advances in SWCNT separations have provided multiple different avenues that generate high-yielding, high-volume samples with single chirality purities of 90% or higher, which has allowed for accelerated progress in the understanding and control of nanotube photophysics [21]. Recently, chemical doping using local chemical functionalization, which is a type of local defect doping, has allowed for the enhancement of the near-infrared (NIR) PL from SWCNTs [19]. Chemical doping changes the crystalline structure of the nanotube, thus shifting the PL of the CNT. For example, oxygen doping in SWCNTs produces a much greater red-shifted and bright PL compared to the original SWCNT PL [21]. Oxygen-doping is performed to regulate the PL properties of SWCNTs by introducing luminescent defects on them. Luminescent defects are deviations in the atomic arrangement that change the periodicity of the luminescence. Oxygen-doped infrared SWCNTs (lf-SWCNTs) exhibit a PL with a high quantum efficiency in the wavelength (photon energy) range of 100–200 nm longer than the intrinsic first sub-band exciton PL peak of the original, pure SWCNTs (**Figure 6**) [19].

A reduced band gap and exciton energy of the dopant-induced local states is the cause of the red-shifted PL feature of the doped nanotubes. The red shift caused by oxygen doping can be clearly seen through brighter PL exhibited by the oxygen-doped If-SWCNT (**Figure 6**).

7.3 Zero-dimensional graphene quantum dots

Graphene quantum dots (GQDs) display some of the most interesting photoluminescent qualities, due to their zero-dimensional (0D) nature. GQDs are a type of carbon dot that are typically less than 100 nm in diameter, a size so small that they are practically 0D, and have an internal graphene lattice [24]. They are a relatively new material; the first synthesis of them was performed by Ponomarenko and Geim in 2008, just 4 years after graphene was first synthesized [17]. Because of their infinitesimally small size, carriers (which become excitons when optically excited)

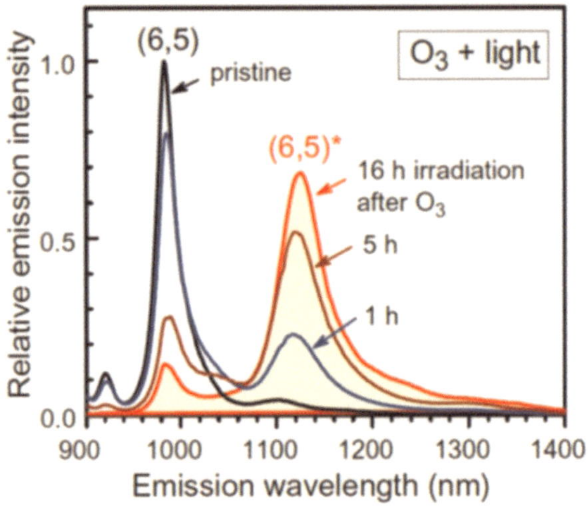

Figure 6.
PL spectral changes of semiconducting (6,5) SWCNTs dispersed in ozone-containing water after UV irradiation. The photon energy of the PL peak labeled (6,5) corresponds to red-shifted PL peak of the doped (6,5) SWCNTs. Permission requested from reference [19].*

within them are said to be confined, since the size of the particle is similar to the Bohr radius of the carrier, which depends on the material [25]. Current quantum confinement theory generalizes that, the smaller the particle, the more blue-shifted its PL spectrum will be [24]. GQDs, however, seem to be an exception to this rule. It has been shown that larger GQDs, on the scale of 60 nm, emit stronger PL in the blue region than smaller GQDs, on the scale of 1.5–5 nm, do.

This strange phenomenon does have an explanation, though, owing to excitons and the unique surface geometry of GQDs. There are two different kinds of excitons that are involved in the PL spectra of GQDs: interior excitons, which are confined in the whole of the QD, and surface excitons, which are confined in edge microstructures on the surface of GQDs. The geometry of GQDs is not as neat as **Figure 7** depicts; rather, the edges of these nanostructures are irregular and take their own shapes, which can change depending on the environment the GQD is in. These edge microstructures have their own localized excitons, which are confined even further than the interior excitons in the body of the GQD. Since these excitons are confined so severely, it follows from quantum confinement theory that they would cause a significant blue shift in the PL spectrum. Larger GQDs are more affected by this simply because they have more surface area than the smaller GQDs, and therefore more edge microstructures that can confine their own excitons [14].

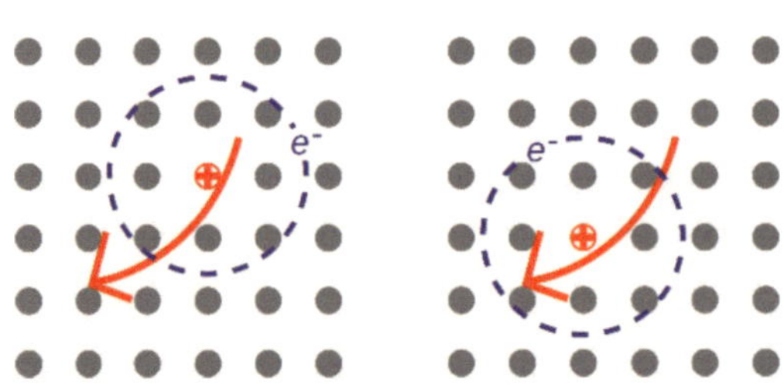

Figure 7.
WM exciton showing movement in a crystal lattice. Image from public domain.

8. Bose-Einstein condensation

Excitons are the quanta of excitation of a medium and are composite particles made of an electron in the conduction band paired with a hole in the valence band. Both the hole and the electron are fermions, meaning they follow the Pauli exclusion principle. However, at sufficiently low density approaching the dilute limit, the particles are spaced far enough to act as a gas of bosons and may undergo a phenomenon known as Bose-Einstein Condensation. At high density and extremely low temperatures, around 0 K to 10 K, this process can occur spontaneously. Bose-Einstein condensation occurs when bosonic atoms or particles have a quantum wavelength on the same order as their spacing. This occurs at temperatures near 0 K and at sufficiently high density, where the spacing approaches the Bohr radius of the particle, upon which the excitons all occupy one quantum state and transition phases into a Bose-Einstein condensate. Then, each exciton has the same energy, and the entire system is one macroscopic quantum wave function. This coherence of state is like the single state of a laser. Bose-Einstein condensation can be used to create spontaneous coherence of a single state in semiconductor, like excitation by a laser. The electron–hole pairs in semiconductors may cohere into a Bose-Einstein condensate at low temperatures, minimizing resistance [26–29].

9. Conclusion

Investigations of exciton systems represent one of the fastest growing, multifaceted, and productive areas of modern solid state physics and material science. Particularly noteworthy in this regard is the deployment of state-of-the-art spectroscopic techniques (e.g., tr-ARPES and femtosecond optical spectroscopy) to better elucidate the structural, dynamical, and quantum mechanical properties of exciton systems.

Author details

John W. Kenney III*, Joshua Jacobsen, Amanda Renfro, Isaac Muñoz and Ruth Christian
Concordia University, Irvine, CA, USA

*Address all correspondence to: john.kenney@cui.edu

References

[1] Knox RS, Robert S. Theory of Excitons. Vol. vii. New York: Academic Press; 1963. p. 207 (Solid state physics; advances in research and applications. Supplement; 5)

[2] Gao B, Hartland GV, Huang L. Transient absorption spectroscopy of excitons in an individual suspended metallic carbon nanotube. Journal of Physical Chemistry Letters. 2013;**4**(18):3050-3055

[3] Guo B, Sun J, Lu Y, Jiang L. Ultrafast dynamics observation during femtosecond laser-material interaction. International Journal of Extreme Manufacturing. 2019;**1**(3):032004

[4] Damascelli A. Probing the electronic structure of complex systems by ARPES. Physica Scripta. 2004;**2004**(T109):61

[5] Sangalli D. Excitons and carriers in transient absorption and time-resolved ARPES spectroscopy: An ab initio approach. Physical Review Materials. 2021;**5**(8):083803

[6] Spano FC, Meskers SCJ, Hennebicq E, Beljonne D. Using circularly polarized luminescence to probe exciton coherence in disordered helical aggregates. The Journal of Chemical Physics. 2008; **129**(2):024704

[7] Kalytchuk S, Zhovtiuk O, Kershaw SV, Zbořil R, Rogach AL. Temperature-dependent exciton and trap-related photoluminescence of CdTe quantum dots embedded in a NaCl matrix: Implication in thermometry. Small. 2016;**12**(4):466-476

[8] Dammak H, Triki S, Mlayah A, Feki H, Abid Y. Excitonic luminescence in the self assembled organic–inorganic quantum well crystal: (C4H3SC2H4–NH3)2[PbI4]. Journal of Luminescence. 2016;**173**:203-207

[9] de Mello DC, Bode M, Meijerink A. Size- and temperature-dependence of exciton lifetimes in CdSe quantum dots. Physical Review B. 2006;**74**(8):085320

[10] Murotani H, Hayakawa Y, Ikeda K, Miyake H, Hiramtsu K, Yamada Y. Temperature dependence of excitonic transitions in Al0.60Ga0.40N/Al0.70Ga0.30N multiple quantum wells from 4 to 750 K. Journal of Applied Physics. 2018;**123**(20):205705

[11] Yang B, Han K. Ultrafast dynamics of self-trapped excitons in lead-free perovskite nanocrystals. Journal of Physical Chemistry Letters. 2021;**12**(34):8256-8262

[12] Fu X, Li F, Lin JF, Gong Y, Huang X, Huang Y, et al. Pressure-dependent light emission of charged and neutral excitons in monolayer MoSe2. Journal of Physical Chemistry Letters. 2017;**8**(15):3556-3563

[13] Wang L, Williams NE, Malachosky EW, Otto JP, Hayes D, Wood RE, et al. Scalable ligand-mediated transport synthesis of organic–inorganic hybrid perovskite nanocrystals with resolved electronic structure and ultrafast dynamics. ACS Nano. 2017; **11**(3):2689-2696

[14] Huang P, Jie SJ, Zhang M, He JX, Xia ZH, Min DY, et al. Anomalous light emission and wide photoluminescence spectra in graphene quantum dot: Quantum confinement from edge microstructure. Journal of Physical Chemistry Letters. 2016;**7**(15):2888-2892

[15] Fang M, Gu H, Song B, Guo Z, Liu S. Thickness and layer stacking order effects on complex optical conductivity

and exciton strength of few-layer graphene: Implications for optical modulators and photodetectors. ACS Applied Nano Materials. 2022;**5**(2): 1864-1872

[16] Gupta S, Johnston A, Khondaker S. Optoelectronic properties of MoS2/ Graphene heterostructures prepared by dry transfer for light-induced energy applications. Journal of Electronic Materials. 16 Jun 2022;**15**(51): 4257-4269

[17] Tian P, Tang L, Teng KS, Lau SP. Graphene quantum dots from chemistry to applications. Materials Today Chemistry. 2018;**10**:221-258

[18] Liu WT, Wu SW, Schuck PJ, Salmeron M, Shen YR, Wang F. Nonlinear broadband photoluminescence of graphene induced by femtosecond laser irradiation. Physical Review B. 2010;**82**(8):081408

[19] Shiraki T, Miyauchi Y, Matsuda K, Nakashima N. Carbon nanotube photoluminescence modulation by local chemical and supramolecular chemical functionalization. Accounts of Chemical Research. 2020;**53**(9):1846-1859

[20] Amori AR, Rossi JE, Landi BJ, Krauss TD. Defects enable dark exciton photoluminescence in single-walled carbon nanotubes. Journal of Physical Chemistry C. 2018;**122**(6): 3599-3607

[21] He X, Htoon H, Doorn SK, Pernice WHP, Pyatkov F, Krupke R, et al. Carbon nanotubes as emerging quantum-light sources. Nature Materials. 2018;**17**(8):663-670

[22] Qin LC. Determination of the chiral indices (n,m) of carbon nanotubes by electron diffraction. Physical Chemistry Chemical Physics. 2006;**9**(1):31-48

[23] Berger S, Voisin C, Cassabois G, Delalande C, Roussignol P, Marie X. Temperature dependence of exciton recombination in semiconducting single-wall carbon nanotubes. Nano Letters. 2007;**7**(2):398-402

[24] Demchenko AP. Excitons in Carbonic Nanostructures. C. 2019;**5**(4): 71

[25] Barbagiovanni EG, Lockwood DJ, Simpson PJ, Goncharova LV. Quantum confinement in Si and Ge nanostructures: Theory and experiment. Applied Physics Reviews. 2014;**1**(1): 011302

[26] Phase Diagram of Degenerate Exciton Systems. [cited 2022 Jul 20]. Available from: https://www.science.org/doi/abs/10.1126/science.1092691

[27] Coherent SD, Waves E. Science. 1996;**273**(5280):1351-1352

[28] Doyle J. Bose–Einstein condensation. Proceedings of the National Academy of Sciences. 1997;**94**(7):2774-2775

[29] Snoke D. Spontaneous bose coherence of excitons and polaritons. Science. 2002;**298**(5597):1368-1372